Engineering Mathematics

Engineering Mathematics

Suleman Ansari

RANDOM PUBLICATIONS
NEW DELHI - 110 002 (INDIA)

Engineering Mathematics

ISBN 978-93-51112-39-6

Published in 2014 in India by

RANDOM PUBLICATIONS

4376-A/4B, Gali Murari Lal, Ansari Road
New Delhi-110 002
Phone: +9111-43580356, 23289044
E-mail: randomexports@gmail.com; sales@randompublications.com; info@randompublications.com

Type Setting by: Friends Media, Delhi-110089
Printed at Thomson Press (India) Ltd

Preface

Historically, engineering mathematics consisted principally of applied analysis, most notably: differential equations; real and complex analysis; approximation theory; Fourier analysis; potential theory; as well as linear algebra and applied probability, outside of analysis. These areas of mathematics were intimately tied to the development of Newtonian physics, and the mathematical physics of that period. This history left a legacy as well: until the early 20th century subjects such as classical mechanics were often taught in applied mathematics departments at American universities, and fluid mechanics may still be taught in (applied) mathematics as well as engineering departments. The success of modern numerical computer methods and software has led to the emergence of computational mathematics, computational science, and computational engineering, which occasionally use high-performance computing for the simulation of phenomena and the solution of problems in the sciences and engineering. These are often considered interdisciplinary fields, but are also of interest to engineering mathematics. Engineering mathematics at tertiary level typically consists of mathematical methods and models courses, and books on the topic are extremely popular. In the English-speaking world, "Engineering mathematics" (or some variant thereof) is the name of several best-selling university textbooks on the subject. In 2013, a search for books containing the phrase 'engineering mathematics' on amazon.com and Google Books, produced more than 192,729 and 457,000 items respectively. However there is some dissension to this viewpoint as to whether or not this is all that constitutes the subject.

Engineering mathematics is a branch of applied mathematics that concerns itself with mathematical methods and techniques that are typically used in engineering and industry. Along with fields like engineering physics and engineering geology, engineering mathematics is an interdisciplinary subject motivated by engineers' needs both for practical, theoretical and other considerations outwith their specialization, and to deal with constraints to be effective in their work.

This book has been designed for senior engineering, mathematics and systems science students. In addition, it has used the optional, advanced sections as the basis for graduate courses in quality control and queueing. It is assumed that the students have taken a first course in probability but that some need a review.

I thank all members of my team who have helped in the preparation of the book. My special thanks go to "Random Publications" who have published the book.

—Suleman Ansari

Contents

1

Systems of Equations and Inequalities

Solving Systems of Equations

Up until this point, we have been dealing with only one equation at a time. Now, we will work with more than variable and more than one equation. These are called systems of equations. When answering a system of equations, you need to give the value for each variable.

Solving Systems of Linear Equations

When we're through covering the two chapters on solving systems of equations, there will be six ways that we can use to solve a system of linear equations

Graphically: Graph both equations and find the intersection point.

Inaccurate by hand.

Useful when using technology.

More appropriate for non-linear systems.

Must solve for the equation for y first.

Substitution: Solve one equation for one variable and then substitute that into the other equation.

Best algebraic technique for non-linear systems.

Works well when a variable can be solved for easily, has a coefficient of one.

Works better when fractions and roots aren't involved.

Addition / Elimination: Multiply one or more equations by a constant and then add the two equations together to eliminate one variable.

Works well for a linear system when there is no variable with a coefficient of one.

Works well for 2x2 (2 equations with 2 variables) systems of equations, but becomes tedious and labour intensive for larger systems.

Gaussian Elimination / Gauss Jordan Elimination: Uses elementary operations to produce equivalent equations.

Works for non-square systems of linear equations.

Built upon the concepts of addition elimination, but instead of obtaining new equations, the old equation is replaced with an equivalent equation.

When applied with matrices, probably the fastest way to solve a large system of linear equations by hand. Certainly the instructor's favourite method.

Cramer's Rule: Uses determinants of a matrix to find the solution.

Works only for square systems of linear equations where the determinant of the coefficient matrix isn't zero.

Good for a computer or calculator where there is a determinant program.

Slow by hand.

Slow on the calculator without a program since every determinant must be entered manually.

Can be used when you need to find just one of the variables.

Matrix Algebra / Matrix Inverses: Uses the inverse of a matrix to find the solution.

Works only for square systems of linear equations where the determinant of the coefficient matrix isn't zero.

Good for a computer or calculator where there is a matrix inverse function.

Slow by hand.

Quick to do on the calculator.

Will return decimal answers, but you can use the fraction key to convert it to integers.

Substitution

The method of substitution will work with non-linear as well as linear equations.

1. Solve one of the equations for one of the variables.
2. Substitute that expression in for the variable in the other equation.
3. Solve the equation for the remaining variable
4. Back-substitute the value for the variable to find the other variable.
5. Check.

The process of back-substitution involves taking the value of the variable found in step 3 and substituting it back into the expression obtained in step 1 (or the original problem) to find the remaining variable. It is important that both variables be given when solving a system of equations.

A common mistake students make is to find one variable and stop there. You need to include a value for all the variables. It is a good idea to check your answer into the both equations, but is probably sufficient to check in the equation you didn't isolate a variable in the first step. That is, if you solved for y in the first equation in step 1, use the second equation to check the answer.

Graphical Approach

The graphical approach works well with a graphing calculator, but is inaccurate by hand (did those points intersect at 1/6 or 1/7?) unless the graph happens to fall exactly on the grid lines.

1. Solve each equation for y. This may involve a plus and minus if there is a y^2 term. If you're not graphing with a calculator or computer, you can skip this step.
2. Graph each equation.
3. Find the points of intersection.
4. Check!

It is important to check your answers to make sure that you have read the intersection point correctly.

Sometimes the calculator will fail to give an intersection point using the intersect command. You may need to use the trace feature of the calculator to find the intersection point. You may use your calculator to check the answer.

Try to convert your answer to fractional form if possible.

The graphical approach can save a lot of time when you're working with a non-linear system of equations.

Systems of Linear Equations in Two Variables

Addition / Elimination

The idea behind the addition / elimination method is to multiple one or more equations by a constant so when they are added together, one of the variables eliminates. Then you have one equation with one variable and you can solve for that variable.

1. Choose a variable to eliminate. Usually the variable that can be eliminated by multiplying by smaller numbers is the better choice.
2. Multiply one or both equations by a constant so that the least common multiple of the coefficients on the variable to be eliminated is obtained. Care should be taken so that one coefficient becomes negative and the other is positive.
3. Add the two equations together so the variable is eliminated.
4. Solve the resulting equation for the remaining variable.
5. Back-substitute that value into the one of the two original equations to find the remaining variable.
6. Check your answer into the other equation.

As an alternative to step 5, and this is extremely helpful when the answer is a fraction or decimal value and not pleasant to work with, you can go through the elimination process again with the other variable, and then you don't have to work with the fractions until the check process.

Graphical Interpretation of Solutions

You may have either no solution, one unique solution, or many solutions when solving a 2x2 system of linear equations.

Exactly One Solution

- Intersecting lines
- Consistent System (consistent means there is a solution, there is no contradiction).
- Independent System (the value for y doesn't depend on what x is).

Write your answer as the set containing the ordered pair {(x,y)} (replace the x and y with the actual values).

This would be analogous to a conditional system from section 2.1, certain values of x and y make it true, while others don't.

Consider the system of linear equations 3x + 2y = 17 and 2x - y = 2. The solution is x = 3 and y = 4, so you write the answer as { (3, 4) }.

No Solution

- Parallel lines
- Inconsistent System (contradiction)
- Independent / Dependent doesn't apply since there is no solution.

Write your answer as "no solution", the symbol for the empty or null set, Ø, or the empty set { }, but do not write it as the set containing the null set {Ø}. Be careful if you're a person who slashes their zeros. Make sure I can tell a zero apart from the null set.

This case occurs when both variables eliminate, and you are left with a false statement.

This would be analogous to a contradiction from section 2.1.

Consider the system of linear equations 3x - 2y = 3 and -3x + 2y = 2. When you add them together, you get 0 = 5, which is a contradiction. Therefore, the answer is "no solution", { }, or Ø

Many Solutions

- Coincident lines (same line)
- Consistent System (identity)
- Dependent System (the value for y will depend on what x is, it's not always the same value).

Write your answer as one of the two equations. Do NOT say "many solutions" or "all real numbers". All real numbers will not work. First of all, it the solutions are ordered pairs, not individual x or y coordinates. Second, not every point works, only those on the line work.

You may also write your answer in parametric form. This will be the preferred method for higher ordered systems, so you might as well learn it now.

This case occurs when both variables eliminate, and you are left with a true statement.

This would be analogous to an identity from section 2.1.

Consider the system of linear equations 3x - 2y = 3 and -3x + 2y = -3. When you add them together, you get 0 = 0, which is always true. Therefore, any values of x and y that satisfy the equation are solutions. You would write your solution as:

- 3x - 2y = 3
- { (x, y) | 3x - 2y = 3 }
- x = t, y = 3/2 t - 3/2 (this is known as parametric form and is covered in section 8.3)

Making "Nice" Problems

Ever wonder why most all of the problems in an algebra textbook come out with nice integer answers? Is it because life is that way? Certainly not, it is because the problems are contrived to have nice answers.

How do you get a made up problem to have nice answers? The answer is that you don't. You start with the answer and work backwards.

Let's say that we want the answer to be (3,-2).

Make up anything for the left side of the equation, say 2x-5y. Then, plug in x=3 and y=-2, so 2(3) - 5(-2) = 6 + 10 = 16. The first equation is 2x-5y=16.

Now, repeat the process again with a different left side, say 3x+2y. Well, 3(3) + 2(-2) = 9 - 4 = 5, so 3x+2y=5.

Your system of linear equations is 2x-5y=16 and 3x+2y=5.

You can make up anything for the left hand side, just make sure the right hand side is what you get when you plug those values in. You will also need to have two equations if there are two variables, one equation if there is only one variable, three equations if there are three variables, etc.

Least Squares Regression Model

Up until this point, when we have found the linear regression model, we have just used the functions on the calculator to obtain the results, and it has been fairly easy and painless. Now we learn that the calculator is actually solving a system of linear equations to obtain the model.

Summation Notation

The capital Greek letter sigma stands for sum. Normally, there is an index with a starting point (k=1) written below the sigma and an ending point (n, meaning k=n) written above the sigma. Then, each variable will have a subscript to let you know that it is a function or sequence that depends on the value of the index, k.

Confused? Well, maybe you should be. The book doesn't discuss sequences and series, they expect you to know how to do it.

I'm going to use a short hand notation to keep things simpler and easier to remember. Instead of writing it the way the book does, I'll just use a sigma symbol and then what I want to total.

$$\sum_{k=1}^{n} x_k = \sum x$$

Remember that sigma means sum, so sigma x means add up all the x's.

In statistics, we like to simplify things, and get a little sloppy, and drop all the index stuff and just know that it applies to all the points. In the notation above, I will use the form on the right. "x just means add up all the x values. Not too bad when you look at it that way.

Linear Regression

Consider the linear model y = ax + b. The values for a and b can be found by solving this system of linear equations.

$$b\Sigma 1 + a\Sigma x = -y$$

$$b\Sigma x + a\Sigma x^2 = -xy$$

Notice each term in the second equation has one more x in it than the corresponding term in the first equation. This pattern will repeat when we do quadratic regression, or cubic, or quartic, or...

You will add up each variable in the summation for each different point. The first summation is the sum of 1. So, if you add 1 for each point, you will simply have the number of points. The other values are the sum of the x's, the sum of the y's, the sum of the squares of the x's, and the sum of the products of the x's and y's.

Writing these ordered pairs in a columnar table, and then adding columns for the x^2 and xy will help. After you solve the system of linear equations, substitute the values for a and b into the equation y = ax + b to get the model.

Find the equation of the linear model that best fits the points (2,3), (5,2), (6,1), and (8,-1).

Set up a table with columns for x, y, x^2, and xy.

x	y	x^2	xy
2	3	4	6
5	2	25	10
6	1	36	6
8	–1	64	-8
21	5	129	14

The numbers in the bottom row represent the sums that go into the system. Since there are 4 points, the –1 = 4. The system of linear equations to solve is 4b + 21a = 5 and 21b + 129a = 14. When you solve that, you get a = –49/75 and b = 117/25. When you stick those back into the model, you get the y = –49/75x + 117/25.

Multivariable Linear Systems

Row-Echelon Form

Row-echelon form is basically a stair-step pattern with leading coefficients of one. When a system is placed into row-echelon form, back substitution is very easy. The bottom row gives the answer for z. That answer is back-substituted into the second equation and y is found. Then both y and z are substituted into the first equation and x is found.

Example: Use back substitution to solve the following system.

$x - y + 2z = 5$

$y - z = -1$

$z = 3$

1. The bottom equation gives us that z = 3.
2. Plugging z = 3 into the second equation gives y – 3 = –1 or y = 2.
3. Plugging y = 2 and z = 3 into the first equation gives
   x – 2 + 2(3) = 5.
4. Solving for x gives 1, so the solution is { (1, 2, 3) }.

Gaussian Elimination

Gaussian Elimination is named after Carl Friedrich Gauss, the German mathematician who proved the fundamental theorem of algebra. Two systems of equations are *equivalent* if they have the same solution set.

Elementary Operations

There are three basic operations, called elementary operations, that can be performed and that will render an equivalent system.

- Interchange two equations.
- Multiply one equation by a non-zero constant.
- Multiply an equation by a non-zero constant and add it to another equation, replacing that equation.

The addition / elimination technique that we discussed in section 5.2 uses these operations, they just weren't formalized. In the elimination technique, we could switch the two equations around and it wouldn't affect the solution set. We could multiply one or both equations by a non-zero constant and then add them together.

There is nothing wrong with multiplying both equations by a non-zero constant, just that what's given here is the elementary or basic operations and multiplying both by a non-zero constant and adding would be a combination of these.

The Number of Solutions to a Linear System

Just like with a 2×2 system of linear equations, a larger system can have one, none, or many solutions.

Unique Solution: A consistent and independent system. The solution is given as an ordered triplet.

No Solution: An inconsistent system. There is no solution, and the answer should be written as the null set or the empty set, but not the set containing the null set.

This results if at any point while trying to solve the system, an equivalent form has a contradiction (variables eliminate and the statement is false).

Many Solutions: A consistent and dependent system. The solution is given in parametric form. This case results when there are less equations than variables and no contradictions.

Example: Solve the following system of linear equations.

$$3x - 2y + z = 1$$

$$y + 2z = 3$$

1. Take the second equation and solve for y to get y = 3 - 2z.
2. Substitute y = 3 – 2z into the first equation for y. This gives you 3x – 2(3 – 2z) + z = 1, which is an equation with two variables, x and z. A little bit of simplification givs 3x – 6 + 4z + z = 1 or 3x + 5z = 7.
3. Since you earlier solved for y in terms of z, we now solve for x in terms of z. It is important that we solve all other variables in terms of the same variable. Solving 3x + 5z = 7 for x gives x = (7 – 5z) / 3
4. Give the solution. Since x and y are both in terms of a third variable, z, we let z = t and get:
   - x = 1/3 (7 – 5t)
   - y = 3 – 2t
   - z = t

Mathematical Models

If you have the right number of ordered pairs, you can fit a model by solving a system a system of linear equations. To have an unique solution, you need as many points as constants.

Linear Model: y=Ax+B

There are two values that need to be found, A and B. Therefore, it takes two points to determine a the equation of a line.

Let's say the line passes through the points (2,3) and (5,7). The resulting system of linear equations, obtained by substituting the values in for x and y is shown. Simply solve the system of linear equations, plug in the values for A and B, and you have the model.

3 = 2A + B

7 = 5A + B

Solving that gives A = 4/3 and B = 1/3. The equation of the line passing through the given points is y = 4/3 x + 1/3.

Quadratic Model: $y=Ax^2 + Bx + C$

This time, there are three variables, A, B, and C. Therefore, it takes three non-collinear points to determine a equation of a parabola.

Let's say the parabola passes through the points (2,3), (5,7), and (8,4). Substitute the values in for x and y and the three linear equations that result is the system that needs to be solved to find A, B, and C.

$3 = 4A + 2B + C$

$7 = 25A + 5B + C$

$4 = 64A + 8B + C$

If you're wondering where there 4, 25, and 64 came from, they are the values of x^2 when $x = 2$, $x = 5$, and $x = 8$.

Solving the system gives $A = -7/18$, $B = 73/18$, and $C = -32/9$. The equation is $y = -7/18\ x^2 + 73/18\ x - 32/9$. This procedure could be extended to find the equation of a cubic passing through four points or a quartic equation passing through five points.

Circle: $x^2 + y^2 + Dx + Ey + F = 0$

You have three variables, D, E, and F. Therefore, it takes three non-collinear points to determine the equation of a circle.

Let's say the circle passes through the points (2,3), (5,7), and (8,4). Substitute the values in for x and y and the equations that result give you the system of linear equations that can give the coefficients that define the circle. You may find it easier to move the constant to the other side.

$4 + 9 + 2D + 3E + F = 0$ becomes $2D + 3E + F = -13$

$25 + 49 + 5D + 7E + F = 0$ becomes $5D + 7E + F = -74$

$64 + 16 + 8D + 4E + F = 0$ becomes $8D + 4E + F = -80$

If you solve that system, you get $D = -69/7$, $E = -55/7$, and $F = 212/7$.

Substituting into the original model gives

$$x^2 + y^2 - 69/7\ x - 55/7\ y + 212/7 = 0.$$

If you were so inclined (this is a pretty ugly problem), you could complete the square to find the centre and radius of the circle.

Least Squares Regression Parabola

If you have exactly three non-collinear points, then the parabola will pass through the points exactly. If there are more than three points, then you must fit a quadratic model to the data, but it won't necessarily pass through all of the points. Recall the least squares regression line that we discussed in the last section that went with the model $y = ax + b$.

$b\Sigma 1 + a\Sigma x = -y$

$b\Sigma x + a\Sigma x^2 = -xy$

Now, we are going to find a system of equations that will fit the model $y = ax^2 + bx + c$. It is an extension of the previous model. Notice it has the same pattern as before.

As you work from left to right on the left hand side, each summation has one additional x in it.

As you work from top to bottom, each term has one more x than the corresponding term in the previous equation.

$$c\Sigma 1 + b\Sigma x + a\Sigma x^2 = -y$$

$$c\Sigma x + b\Sigma x^2 + a\Sigma x^3 = -xy$$

$$c\Sigma x^2 + b\Sigma x^3 + a\Sigma x^4 = -x^2y$$

If you wanted to find a least squares regression cubic, then you would need four variables and at least five points (four points would exactly fit the model).

Follow the same pattern as before, adding one more variable (for d) and one more row.

Partial Fractions

Addition of Rational Expressions

In arithmetic, you learned how to add fractions. You found the least common denominator, and then multiplied both the numerator and denominator of each term by what was needed to complete the common denominator.

$$\frac{1}{x-1}+\frac{2x-1}{x^2+3}=\left(\frac{1}{x-1}\right)\left(\frac{x^2+3}{x^2+3}\right)+\left(\frac{2x-1}{x^2+3}\right)\left(\frac{x-1}{x-1}\right)$$

$$=\frac{x^2+3+2x^2-3x+1}{(x-1)(x^2+3)}$$

$$=\frac{3x^2-3x+4}{(x-1)(x^2+3)}$$

In algebra, you have carried that process on to addition of rational expressions.

You once again multiplied the numerator and denominator of each term by what was missing from the denominator of that term.

With Partial Fraction Decomposition, we're going to reverse the process and decompose a rational expression into two or more simpler proper rational expressions that were added together.

Partial Fraction Decomposition

Partial Fraction Decomposition only works for proper rational expressions, that is, the degree of the numerator must be less than the degree of the denominator. If it is not, then you must perform long division first, and then perform the partial fraction decomposition on the rational part (the remainder over the divisor).

After you've done the partial fraction decomposition, just add back in the quotient part from the long division. When discussing polynomials in section 3.4, we learned that every polynomial with real coefficients can be factored using only linear and irreducible quadratic factors. This means that there are only two types of factors that we have to worry about.

Linear Factors

If the partial fractions we're decomposing the rational expression into must be proper, then the only thing that can be over a linear factor is a constant. So, for every linear factor in the denominator, you will need a constant in over that in the numerator.

Irreducible Quadratic Factors.

If the partial fractions we're decomposing the rational expression into must be proper, then an irreducible quadratic factor could have a linear term and/or a constant term in the numerator. So, for every irreducible quadratic factor in the denominator, you will need a linear term and a constant term in the numerator.

Repeated Factors

Consider a fraction in which the denominator is 8. Does that mean that the denominator of each term being added together had to be an 8? No, the denominators could have been 2, 4, or 8 because the common denominator between 2, 4, and 8 is 8. The implications of this for partial fraction decomposition are that when you have a repeated factor (a factor with a multiplicity other than one), you need to include a factor in the expansion for each power possible.

For example, if you have an $(x-2)^3$, you will need to include an $(x-2)$, an $(x-2)^2$, and an $(x-2)^3$.

The exponents of 2 or 3 does not change whether the factor is linear or quadratic, only how many times the factor is there. Each of those $(x-2)$ factors would receive a constant term in the numerator because $x-2$ is linear, no matter what power it is raised to.

Improper Fractions

A proper fraction is a fraction where the numerator is less than the denominator. For rational expressions, that means the degree in the numerator is less than the degree in the denominator.

If you have an improper fraction, you first have to perform long division to get a quotient and a remainder.

Leave the quotient part alone, but perform the partial fraction decomposition on the fraction that remained.

The Basic Equation

After setting up the partial fraction equation, you multiply both sides of the equation by the least common denominator to get rid of the fractions. The resulting equation, without the fractions, is called the Basic Equation.

Technique 1 - Pick Values for x

$$\frac{x+2}{x(x-4)} = \frac{A}{x} + \frac{B}{x-4}$$

When there are linear factors, the easiest way to do the decomposition is probably to pick nice values for x.

In the example to the right, we take each factor in the denominator and give it its own term on the right side. Since each factor in the denominator is linear and the rational expressions need to be proper, a constant term was placed in each numerator of each term.

We multiply through by the least common denominator to arrive at the equation without the fractions. This equation is called the Basic Equation and is...

$$x + 2 = A(x - 4) + Bx$$

"Nice" values for x are those that cause each linear factor to be zero. In this case, x=4 and x=0 are nice. "Nice" values will cause every term except for one to drop out of the equation, and so you will be able to find the value of a variable very quickly.

When you let x = 0, the B term drops out and you get...

$0 + 2 = A(0 - 4)$ or $2 = -4A$. Solving that gives $A = -1/2$.

When you let x = 4, the A term drops out and you get...

$4 + 2 = B(4)$ or $6 = 4B$. Solving that gives $B = 3/2$.

$$\frac{x+2}{x(x-4)} = \frac{-1/2}{x} + \frac{3/2}{x-4}$$
$$= \frac{1}{2}\left(\frac{3}{x-4} - \frac{1}{x}\right)$$

After you have found the values of each constant, it is important that you plug those values back into the decomposition. Don't just stop with A=-1/2, B=3/2 because someone else may have defined A and B differently.

The correct answer is to place them back into the decomposition and simplify if necessary. The simplification would be to reduce the number of signs (don't have the first term negative if possible) and to eliminate the compound fractions by factoring out the least common denominator of the numerators.

Note: There are only as many nice values as there are distinct (different) linear factors. If there are repeated linear factors or irreducible quadratic factors (repeated or not), you will not have enough "nice" values to pick.

In cases like that, you will have to pick convenient but not-so-nice values and then substitute the known constants into the equation to find the other constants. Plug in easy numbers like x=0, x=1, etc.

You will need to pick as many values for x as there are constants to be found.

Technique 2 - Creating a System of Linear Equations

The first technique of picking values for x works extremely well when all the factors are distinct linear factors. If there are any linear factors, then first method is still probably the better technique to use. However, if there are only irreducible quadratic factors, then the method of picking values for x can become messy.

$$\frac{2x^2+x+8}{(x^2+4)^2} = \frac{Ax+B}{x^2+4} + \frac{Cx+D}{(x^2+4)^2}$$

There is another way to do these problems (actually, this technique will work when there are linear factors, just that the other is easier and faster). The first part of the process is the same. Go ahead and write out the decomposition, including constant terms over linear factors and linear and constant terms over the irreducible quadratic factors. Then, multiply through by the least common denominator to find the basic equation. That much is the same. The basic equation is...

$$2x^2 + x + 8 = Ax(x^2 + 4) + B(x^2+4) + Cx + D$$

Now the difference comes.

Go ahead and expand (multiply out) the basic equation...

$$2x^2 + x + 8 = Ax^3 + 4Ax + Bx^2 + 4B + Cx + D$$

and regroup the terms by common powers of the variable x.

$$2x^2 + x + 8 = Ax^3 + Bx^2 + 4Ax + Cx + 4B + D$$

Now factor out the x's by power.

$$2x^2 + x + 8 = (A)x^3 + (B)x^2 + (4A + C)x + (4B + D) \quad (1)$$

The next part works because if two polynomials are going to be equal, they must have the same number of similar terms on both sides. So the trick here is to equate both sides of the equation together by equating similar terms.

Set the x^3 terms on the left (of which there are none) equal to the x^3 terms on the right (of which there are A) to arrive at the first equation 0=A. Well, that was pretty easy, and already you have the value for A.

Set the x^2 terms on the left (of which there are 2) equal to the x^2 terms on the right (of which there are B) to arrive at the second equation of 2=B. Now you know B.

For the x's, there is 1 on the left and 4A+C on the right, so 1=4A+C.

For the constants, there are 8 on the left and 4B+D on the right, so 8=4B+D.

You may have to solve a system of linear equations, but that's why this section on partial fractions is in this chapter. We just refreshed our memories on how to solve a system of linear equations in the last section.

$$\frac{2x^2 + x + 8}{(x^2 + 4)^2} = \frac{0x + 2}{x^2 + 4} + \frac{1x + 0}{(x^2 + 4)^2}$$

$$= \frac{2}{x^2 + 4} + \frac{x}{(x^2 + 4)^2}$$

After some solving, we arrive at A=0, B=2, C=1, and D=0 and we are ready to plug those values back into the original decomposition to arrive at the final answer.

Be sure and simplify the answer if necessary. This example was actually pretty easy because you were able to find A and B right off the bat. Many times, you will need to solve a much larger system of equations to find the values.

Systems of Inequalities

Graph of an Inequality

When we were dealing with one variable, the graph of an inequality was some interval on the real number line. Now we are dealing with two variables, and the graph of an inequality is the set of all ordered pairs (x, y) that satisfy the inequality. Typically, this will be represented by a half-plane. Sketching an Inequality in Two Variables:

1. Replace the inequality with an equal sign and sketch the resulting equation. Use a solid line if the equal to is included (<= or >=) and a dashed line of the inequality is strict (< or >).
2. Pick a test point not on the graph of the equation. If that test point satisfies the inequality, then shade every point on the side of the inequality that includes your test point. If the test point doesn't satisfy the inequality, then shade the other side of the inequality.

Figure: *Solving a System of Inequalities*

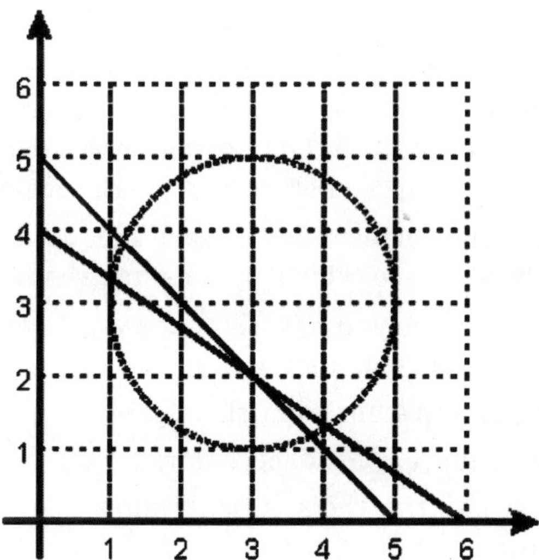

The solution to a system of inequalities is the set of all ordered pairs (x, y) that satisfy all the inequalities simultaneously.

The solution to the system of linear inequalities given is sketched to the right.

$(x-3)^2+(y-3)^2<4$

$x+y<=5$

$2x+3y<=12$

Notice the circle is dotted (dashed) instead of solid. This is because there is a strict inequality (< or >), that is, the equal to is not included.

Now, one misleading thing about generating the output on a computer is that it looks very easy to sketch the system. In reality, there are four ways you can sketch the solution to a system of linear inequalities

1. Draw all the lines and curves. Pick a test point in each region (there will be many regions - there are 8 in the figure shown). Drawback - way too much work.
2. Shade each inequality as it is encountered. Whatever is shaded for each inequality is the solution. If there are three inequalities, whatever is shaded three times is the solution. If there are five inequalities, then whatever is shaded five times is the solution. Drawback - if you do a good job of shading, and you should, it becomes difficult to tell what is shaded three times as opposed to what is shaded four times.
3. Shade each inequality as it is encountered, but shade the false part in each case. The contrapositive of "if doesn't work, then shaded" is "if not shaded, then works". In other words, whatever is left unshaded will be the part that is the solution. Drawback - your answers will be unshaded where the books will be shaded.
4. Shade just the correct edge of each inequality. Then you can extend those to see which regions are shaded on all sides.

Whichever technique you use, right the word "yes" or something similar in the solution region.

Determining the System from the Graph.

1. Write the equation for each boundary on the graph.
2. Pick a test point in the solution region and substitute it into the equation.

3. Replace the equal sign by whichever inequality is needed to make the statement true. Include the equal sign if it is a solid line and don't include the equal sign in the inequality if it is a dashed line.

Linear Programming

In business, it is often desirable to find the production levels that will produce the maximum profit or the minimum cost. The production process can often be described with a set of linear inequalities called constraints. The profit or cost function to be maximized or minimized is called the objective function. The process of finding the optimal levels with the system of linear inequalities is called linear programming (as opposed to non-linear programming).

Definitions

Objective Function: The linear function (equal sign) representing cost, profit, or some other quantity to be maximized of minimized subject to the constraints.

Constraints: A system of linear inequalities.

Problem Constraints: The linear inequalities that are derived from the application. For example, there may be only 40 hours a machine can be used in a week, so the total time it is used would have to be <= 40. The problem constraints are usually stated in the story problem.

Non-Negativity Constraints

The linear inequalities x>=0 and y>=0. These are included because x and y are usually the number of items produced and you cannot produce a negative number of items, the smallest number of items you could produce is zero.

These are not (usually) stated, they are implied.

Feasible Region

The solution to the system of linear inequalities. That is, the set of all points that satisfy all the constraints. Only points in the feasible region can be used.

Corner Point

A vertex of the feasible region. Not every intersection of lines is a corner point. The corner points only occur at a vertex of the feasible region. If there is going to be an optimal solution to a linear programming problem, it will occur at one or more corner points, or on a line segment between two corner points.

Bounded Region

A feasible region that can be enclosed in a circle. A bounded region will have both a maximum and minimum values.

Unbounded Region

A feasible region that can not be enclosed in a circle.

Fundamental Theorem of Linear Programming: Recall that almost every area of mathematics has its fundamental theorem. Here are some of the fundamental theorems or principles that occur in your text.

Fundamental Theorem of Arithmetic: Every integer greater than one is either prime or can be expressed as an unique product of prime numbers.

Fundamental Theorem of Algebra : Every polynomial in one variable of degree $n > 0$ has at least one real or complex zero.

Fundamental Counting Principle : If there are m ways to do one thing, and n ways to do another, then there are m*n ways of doing both.

Fundamental Theorem of Linear Programming

If there is a solution to a linear programming problem, then it will occur at a corner point, or on a line segment between two corner points.

The Fundamental Theorem of Linear Programming is a great help. Instead of testing all of the infinite number of points in the feasible region, you only have to test the corner points. Whichever corner point yields the largest value for the objective function is the maximum and whichever corner point yields the smallest value for the objective function is the minimum.

Solving a Linear Programming Problem

If the problem is not a story problem, skip to step 3.

1. Define the variables. Usually, a good choice for the definition is the quantity they asked you to find in the problem.
2. Write the problem by defining the objective function and the system of linear inequalities. Don't forget about the non-negativity constraints where necessary.

3. Sketch the system of linear inequalities to obtain the feasible region.
4. Identify each corner point of the feasible region. You can find the corner points by forming a 2x2 system of linear equations from the two lines that intersect at that point and solving that system.
5. Evaluate the objective function at each corner point.
6. Choose the point yielding the largest value or smallest value depending on whether the problem is a maximization or minimization problem.

Be careful how you give the answer. The answer should give not only the maximum or minimum value (the value of the objective function), but it should also give the location where that extremum occurs. Example:

The maximum value is 9 when x=2 and y=3. If it is a story problem, then give the answer in terms of the original definitions of x and y.

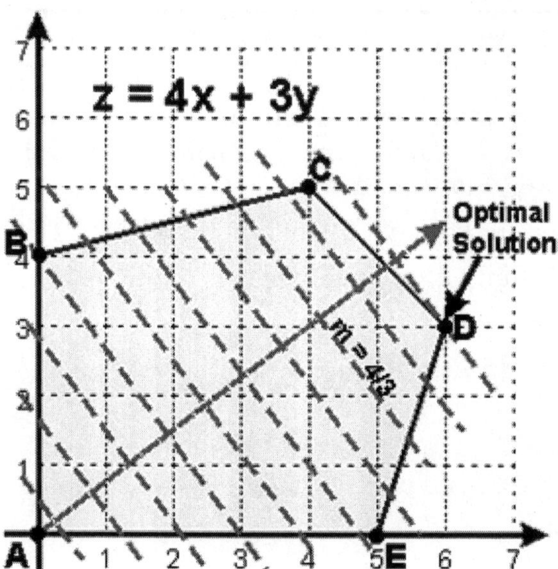

Figure: *Geometric Approach*

If the slope of the objective function is negative and you take a line with that slope passing through the origin and move it to the right through the feasible region, the last corner point hit by that moving line will be the maximum value.

In the example shown, the last line with slope m=-4/3 that touches the feasible region touches at the corner point (6,3).

Since z=4(6)+3(3)=24+9=33, the maximum value is 33 when x=6 and y=3.

Algebraic Approach

Now, to verify the solution non-geometrically. Since we know the optimal solution has to occur at one or more corner points, we make a table listing all the corner points and evaluate the objective function at those points.

Corner Point	x	y	z = 4x + 3y
A	0	0	0
B	0	4	12
C	4	5	31
D	6	3	33
E	5	0	20

As you can see, the corner point with the maximum value is at (6,3).

We can also determine the minimum value from that table. A suitable answer, assuming the problem had asked for both the maximum and minimum is...

The minimum value is 0 when x=0 and y=0.

The maximum value is 33 when x=6 and y=3.

2

Mathematical Logic

Mathematical logic is a subfield of mathematics with close connections to computer science and philosophical logic. The field includes the mathematical study of logic and the applications of formal logic to other areas of mathematics. The unifying themes in mathematical logic include the study of the expressive power of formal systems and the deductive power of formal proof systems.

Mathematical logic is often divided into the subfields of set theory, model theory, recursion theory, and proof theory and constructive mathematics. These areas share basic results on logic, particularly first-order logic, and definability.

Since its inception, mathematical logic has contributed to, and has been motivated by, the study of foundations of mathematics. This study began in the late 19th century with the development of axiomatic frameworks for geometry, arithmetic, and analysis. In the early 20th century it was shaped by David Hilbert's program to prove the consistency of foundational theories.

Results of Kurt Gödel, Gerhard Gentzen, and others provided partial resolution to the program, and clarified the issues involved in proving consistency. Work in set theory showed that almost all ordinary mathematics can be formalized in terms of sets, although there are some theorems that cannot be proven in common axiom systems for set theory. Contemporary work in the foundations of mathematics often focuses on establishing which parts of mathematics can be formalized in particular formal systems, rather than trying to find theories in which all of mathematics can be developed.

History

Mathematical logic emerged in the mid-19th century as a subfield of mathematics independent of the traditional study of logic (Ferreirós 2001, p. 443). Before this emergence, logic was studied with rhetoric, through the syllogism, and with philosophy. The first half of the 20th century saw an explosion of fundamental results, accompanied by vigorous debate over the foundations of mathematics.

Early History

Sophisticated theories of logic were developed in many cultures, including China, India, Greece and the Islamic world. In the 18th century, attempts to treat the operations of formal logic in a symbolic or algebraic way had been made by philosophical mathematicians including Leibniz and Lambert, but their labours remained isolated and little known.

19th Century

In the middle of the nineteenth century, George Boole and then Augustus De Morgan presented systematic mathematical treatments of logic. Their work, building on work by algebraists such as George Peacock, extended the traditional Aristotelian doctrine of logic into a sufficient framework for the study of foundations of mathematics .

Charles Sanders Peirce built upon the work of Boole to develop a logical system for relations and quantifiers, which he published in several papers from 1870 to 1885. Gottlob Frege presented an independent development of logic with quantifiers in his *Begriffsschrift*, published in 1879. Frege's work remained obscure, however, until Bertrand Russell began to promote it near the turn of the century. The two-dimensional notation Frege developed was never widely adopted and is unused in contemporary texts.

From 1890 to 1905, Ernst Schröder published *Vorlesungen über die Algebra der Logik* in three volumes. This work summarized and extended the work of Boole, De Morgan, and Peirce, and was a comprehensive reference to symbolic logic as it was understood at the end of the 19th century.

Foundational Theories

Some concerns that mathematics had not been built on a proper foundation led to the development of axiomatic systems for fundamental areas of mathematics such as arithmetic, analysis, and geometry.

In logic, the term *arithmetic* refers to the theory of the natural numbers. Giuseppe Peano (1888) published a set of axioms for arithmetic that came to bear his name (Peano axioms), using a variation of the logical system of Boole and Schröder but adding quantifiers. Peano was unaware of Frege's work at the time. Around the same time Richard Dedekind showed that the natural numbers are uniquely characterized by their induction properties. Dedekind (1888) proposed a different characterization, which lacked the formal logical character of Peano's axioms. Dedekind's work, however, proved theorems inaccessible in Peano's system, including the uniqueness of the set of natural numbers (up to isomorphism) and the recursive definitions of addition and multiplication from the successor function and mathematical induction.

In the mid-19th century, flaws in Euclid's axioms for geometry became known. In addition to the independence of the parallel postulate, established by Nikolai Lobachevsky in 1826, mathematicians discovered that certain theorems taken for granted by Euclid were not in fact provable from his axioms. Among these is the theorem that a line contains at least two points, or that circles of the same radius whose centres are separated by that radius must intersect. Hilbert (1899) developed a complete set of axioms for geometry, building on previous work by Pasch (1882). The success in axiomatizing geometry motivated Hilbert to seek complete axiomatizations of other areas of mathematics, such as the natural numbers and the real line. This would prove to be a major area of research in the first half of the 20th century.

The 19th century saw great advances in the theory of real analysis, including theories of convergence of functions and Fourier series. Mathematicians such as Karl Weierstrass began to construct functions that stretched intuition, such as nowhere-differentiable continuous functions. Previous conceptions of a function as a rule for computation, or a smooth graph, were no longer adequate. Weierstrass began to advocate the arithmetization of analysis, which sought to axiomatize analysis using properties of the natural numbers. The modern "ε-δ" definition of limits and continuous functions was developed by Bolzano and Cauchy between 1817 and 1823. In 1858, Dedekind proposed a definition of the real numbers in terms of Dedekind cuts of rational numbers, a definition still employed in contemporary texts.

Georg Cantor developed the fundamental concepts of infinite set theory. His early results developed the theory of cardinality and proved that the reals and the natural numbers have different cardinalities. Over the next twenty years, Cantor developed a theory of transfinite numbers in a series of publications. In 1891, he published a new proof of the uncountability of the real numbers that introduced the diagonal argument, and used this method to prove Cantor's theorem that no set can have the same cardinality as its powerset. Cantor believed that every set could be well-ordered, but was unable to produce a proof for this result, leaving it as an open problem in 1895.

20th Century

In the early decades of the 20th century, the main areas of study were set theory and formal logic. The discovery of paradoxes in informal set theory caused some to wonder whether mathematics itself is inconsistent, and to look for proofs of consistency.

In 1900, Hilbert posed a famous list of 23 problems for the next century. The first two of these were to resolve the continuum hypothesis and prove the consistency of elementary arithmetic, respectively; the tenth was to produce a method that could decide whether a multivariate polynomial equation over the integers has a solution. Subsequent work to resolve these problems shaped the direction of mathematical logic, as did the effort to resolve Hilbert's *Entscheidungsproblem*, posed in 1928. This problem asked for a procedure that would decide, given a formalized mathematical statement, whether the statement is true or false.

Set Theory and Paradoxes

Ernst Zermelo (1904) gave a proof that every set could be well-ordered, a result Georg Cantor had been unable to obtain. To achieve the proof, Zermelo introduced the axiom of choice, which drew heated debate and research among mathematicians and the pioneers of set theory. The immediate criticism of the method led Zermelo to publish a second exposition of his result, directly addressing criticisms of his proof. This paper led to the general acceptance of the axiom of choice in the mathematics community.

Skepticism about the axiom of choice was reinforced by recently discovered paradoxes in naive set theory. Cesare Burali-Forti (1897) was the first to state a paradox: the Burali-Forti paradox shows that the collection of all ordinal numbers cannot form a set. Very soon thereafter, Bertrand Russell discovered Russell's paradox in 1901, and Jules Richard (1905) discovered Richard's paradox.

Zermelo (1908) provided the first set of axioms for set theory. These axioms, together with the additional axiom of replacement proposed by Abraham Fraenkel, are now called Zermelo–Fraenkel set theory (ZF). Zermelo's axioms incorporated the principle of limitation of size to avoid Russell's paradox.

In 1910, the first volume of *Principia Mathematica* by Russell and Alfred North Whitehead was published. This seminal work developed the theory of functions and cardinality in a completely formal framework of type theory, which Russell and Whitehead developed in an effort to avoid the paradoxes. *Principia Mathematica* is considered one of the most influential works of the 20th century, although the framework of type theory did not prove popular as a foundational theory for mathematics.

Fraenkel (1922) proved that the axiom of choice cannot be proved from the remaining axioms of Zermelo's set theory with urelements. Later work by Paul Cohen (1966) showed that the addition of urelements is not needed, and the axiom of choice is unprovable in ZF. Cohen's proof developed the method of forcing, which is now an important tool for establishing independence results in set theory.

Symbolic Logic

Leopold Löwenheim (1918) and Thoralf Skolem (1919) obtained the Löwenheim–Skolem theorem, which says that first-order logic cannot control the cardinalities of infinite structures. Skolem realized that this theorem would apply to first-order formalizations of set theory, and that it implies any such formalization has a countable model. This counterintuitive fact became known as Skolem's paradox.

In his doctoral thesis, Kurt Gödel (1929) proved the completeness theorem, which establishes a correspondence between syntax and semantics in first-order logic. Gödel used the completeness theorem to prove the compactness theorem, demonstrating the finitary nature of first-order logical consequence. These results helped establish first-order logic as the dominant logic used by mathematicians.

In 1931, Gödel published *On Formally Undecidable Propositions of Principia Mathematica and Related Systems*, which proved the incompleteness (in a different meaning of the word) of all sufficiently strong, effective first-order theories. This result, known as Gödel's incompleteness theorem, establishes severe limitations on axiomatic foundations for mathematics, striking a strong blow to Hilbert's program. It showed the impossibility of providing a consistency proof of arithmetic within any formal theory of arithmetic. Hilbert, however, did not acknowledge the importance of the incompleteness theorem for some time.

Gödel's theorem shows that a consistency proof of any sufficiently strong, effective axiom system cannot be obtained in the system itself, if the system is consistent, nor in any weaker system. This leaves open the possibility of consistency proofs that cannot be formalized within the system they consider. Gentzen (1936) proved the consistency of arithmetic using a finitistic system together with a principle of transfinite induction.

Gentzen's result introduced the ideas of cut elimination and proof-theoretic ordinals, which became key tools in proof theory. Gödel (1958) gave a different consistency proof, which reduces the consistency of classical arithmetic to that of intutitionistic arithmetic in higher types.

Beginnings of the Other Branches

Alfred Tarski developed the basics of model theory.

Beginning in 1935, a group of prominent mathematicians collaborated under the pseudonym Nicolas Bourbaki to publish a series of encyclopedic mathematics texts. These texts, written in an austere and axiomatic style, emphasized rigorous presentation and set-theoretic foundations. Terminology coined by these texts, such as the words *bijection*, *injection*, and *surjection*, and the set-theoretic foundations the texts employed, were widely adopted throughout mathematics.

The study of computability came to be known as recursion theory, because early formalizations by Gödel and Kleene relied on recursive definitions of functions. When these definitions were shown equivalent to Turing's formalization involving Turing machines, it became clear that a new concept – the computable function – had been discovered, and that this definition was robust enough to admit numerous independent characterizations.

In his work on the incompleteness theorems in 1931, Gödel lacked a rigorous concept of an effective formal system; he immediately realized that the new definitions of computability could be used for this purpose, allowing him to state the incompleteness theorems in generality that could only be implied in the original paper.

Numerous results in recursion theory were obtained in the 1940s by Stephen Cole Kleene and Emil Leon Post. Kleene (1943) introduced the concepts of relative computability, foreshadowed by Turing (1939), and the arithmetical hierarchy. Kleene later generalized recursion theory to higher-order functionals. Kleene and Kreisel studied formal versions of intuitionistic mathematics, particularly in the context of proof theory.

Subfields and Scope

Contemporary mathematical logic is roughly divided into four areas: set theory, model theory, recursion theory, and proof theory and constructive mathematics. Each area has a distinct focus, although many techniques and results are shared between multiple areas. The border lines between these fields, and the lines between mathematical logic and other fields of mathematics, are not always sharp. Gödel's incompleteness theorem marks not only a milestone in recursion theory and proof theory, but has also led to Loeb's theorem in modal logic. The method of forcing is employed in set theory, model theory, and recursion theory, as well as in the study of intuitionistic mathematics.

The mathematical field of category theory uses many formal axiomatic methods, and includes the study of categorical logic, but category theory is not ordinarily considered a subfield of mathematical logic. Because of its applicability in diverse fields of mathematics, mathematicians including Saunders Mac Lane have proposed category theory as a foundational system for mathematics, independent of set theory. These foundations use toposes, which resemble generalized models of set theory that may employ classical or nonclassical logic.

Formal Logic

At its core, mathematical logic deals with mathematical concepts expressed using formal logical systems. These systems, though they differ in many details, share the common property of considering only expressions in a fixed formal language, or signature. The system of first-order logic is the most widely studied today, because of its applicability to foundations of mathematics and because of its desirable proof-theoretic properties. Stronger classical logics such as second-order logic or infinitary logic are also studied, along with nonclassical logics such as intuitionistic logic.

First-order Logic

First-order logic is a particular formal system of logic. Its syntax involves only finite expressions as well-formed formulas, while its semantics are characterized by the limitation of all quantifiers to a fixed domain of discourse.

Early results about formal logic established limitations of first-order logic. The Löwenheim–Skolem theorem (1919) showed that if a set of sentences in a countable first-order language has an infinite model then it has at least one model of each infinite cardinality. This shows that it is impossible for a set of first-order axioms to characterize the natural numbers, the real numbers, or any other infinite structure up to isomorphism. As the goal of early foundational studies was to produce axiomatic theories for all parts of mathematics, this limitation was particularly stark.

Gödel's completeness theorem established the equivalence between semantic and syntactic definitions of logical consequence in first-order logic. It shows that if a particular sentence is true in every model that satisfies a particular set of axioms, then there must be a finite deduction of the sentence from the axioms. The compactness theorem first appeared as a lemma in Gödel's proof of the completeness theorem, and it took many years before logicians grasped its significance and began to apply it routinely. It says that a set of sentences has a model if and only if every finite subset has a model, or in other words that an inconsistent set of formulas must have a finite inconsistent subset. The completeness and compactness theorems allow for sophisticated analysis of logical consequence in first-order logic and the development of model theory, and they are a key reason for the prominence of first-order logic in mathematics.

Gödel's incompleteness theorems establish additional limits on first-order axiomatizations. The first incompleteness theorem states that no sufficiently strong, effectively given logical system can prove its own consistency unless it is actually inconsistent.

Here a logical system is effectively given if it is possible to decide, given any formula in the language of the system, whether the formula is an axiom. When applied to first-order logic, the first incompleteness theorem implies that any sufficiently strong, consistent, effective first-order theory has models that are not elementarily equivalent, a stronger limitation than the one established by the Löwenheim–

Skolem theorem. The second incompleteness theorem states that no sufficiently strong, consistent, effective axiom system for arithmetic can prove its own consistency, which has been interpreted to show that Hilbert's program cannot be completed.

Other Classical Logics

Many logics besides first-order logic are studied. These include infinitary logics, which allow for formulas to provide an infinite amount of information, and higher-order logics, which include a portion of set theory directly in their semantics.

The most well studied infinitary logic is $L\omega_2,\omega$. In this logic, quantifiers may only be nested to finite depths, as in first order logic, but formulas may have finite or countably infinite conjunctions and disjunctions within them. Thus, for example, it is possible to say that an object is a whole number using a formula of such as

$$(x = 0) \vee (x = 1) \vee (x = 2) \vee \ldots.$$

Higher-order logics allow for quantification not only of elements of the domain of discourse, but subsets of the domain of discourse, sets of such subsets, and other objects of higher type. The semantics are defined so that, rather than having a separate domain for each higher-type quantifier to range over, the quantifiers instead range over all objects of the appropriate type. The logics studied before the development of first-order logic, for example Frege's logic, had similar set-theoretic aspects. Although higher-order logics are more expressive, allowing complete axiomatizations of structures such as the natural numbers, they do not satisfy analogues of the completeness and compactness theorems from first-order logic, and are thus less amenable to proof-theoretic analysis.

Nonclassical and Modal Logic

Modal logics include additional modal operators, such as an operator which states that a particular formula is not only true, but necessarily true. Although modal logic is not often used to axiomatize mathematics, it has been used to study the properties of first-order provability and set-theoretic forcing.

Intuitionistic logic was developed by Heyting to study Brouwer's program of intuitionism, in which Brouwer himself avoided formalization. Intuitionistic logic specifically does not include the law of the excluded middle, which states that each sentence is either true or its negation is true. Kleene's work with the proof theory of intuitionistic logic showed that constructive information can be recovered from intuitionistic proofs. For example, any provably total function in intuitionistic arithmetic is computable; this is not true in classical theories of arithmetic such as Peano arithmetic.

Algebraic Logic

Algebraic logic uses the methods of abstract algebra to study the semantics of formal logics. A fundamental example is the use of Boolean algebras to represent truth values in classical propositional logic, and the use of Heyting algebras to represent truth values in intuitionistic propositional logic. Stronger logics, such as first-order logic and higher-order logic, are studied using more complicated algebraic structures such as cylindric algebras.

Set Theory

Set theory is the study of sets, which are abstract collections of objects. Many of the basic notions, such as ordinal and cardinal numbers, were developed informally by Cantor before formal axiomatizations of set theory were developed. The first such axiomatization, due to Zermelo (1908), was extended slightly to become Zermelo–Fraenkel set theory (ZF), which is now the most widely-used foundational theory for mathematics.

Other formalizations of set theory have been proposed, including von Neumann–Bernays–Gödel set theory (NBG), Morse–Kelley set theory (MK), and New Foundations (NF). Of these, ZF, NBG, and MK are similar in describing a cumulative hierarchy of sets. New Foundations takes a different approach; it allows objects such as the set of all sets at the cost of restrictions on its set-existence axioms. The system of Kripke-Platek set theory is closely related to generalized recursion theory.

Two famous statements in set theory are the axiom of choice and the continuum hypothesis. The axiom of choice, first stated by Zermelo (1904), was proved independent of ZF by Fraenkel (1922), but has come to be widely accepted by mathematicians. It states that given a collection of nonempty sets there is a single set C that contains exactly one element from each set in the collection. The set C is said to "choose" one element from each set in the collection. While the ability to make such a choice is considered obvious by some, since each set in the collection is nonempty, the lack of a general, concrete rule by which the choice can be made renders the axiom nonconstructive. Stefan Banach and Alfred Tarski (1924) showed that the axiom of choice can be used to decompose a solid ball into a finite number of pieces which can then be rearranged, with no scaling, to make two solid balls of the original size. This theorem, known as the Banach-Tarski paradox, is one of many counterintuitive results of the axiom of choice.

The continuum hypothesis, first proposed as a conjecture by Cantor, was listed by David Hilbert as one of his 23 problems in 1900. Gödel showed that the continuum hypothesis cannot be disproven from the axioms of Zermelo-Frankel set theory (with or without the axiom of choice), by developing the constructible universe of set theory in which the continuum hypothesis must hold.

In 1963, Paul Cohen showed that the continuum hypothesis cannot be proven from the axioms of Zermelo-Frankel set theory (Cohen 1966). This independence result did not completely settle Hilbert's question, however, as it is possible that new axioms for set theory could resolve the hypothesis. Recent work along these lines has been conducted by W. Hugh Woodin, although its importance is not yet clear.

Contemporary research in set theory includes the study of large cardinals and determinacy. Large cardinals are cardinal numbers with particular properties so strong that the existence of such cardinals cannot be proved in ZFC. The existence of the smallest large cardinal typically studied, an inaccessible cardinal, already implies the consistency of ZFC. Despite the fact that large cardinals have extremely high cardinality, their existence has many ramifications for the structure of the real line. *Determinacy* refers to the possible existence of winning strategies for certain two-player games (the games are said to be *determined*). The existence of these strategies implies structural properties of the real line and other Polish spaces.

Model Theory

Model theory studies the models of various formal theories. Here a theory is a set of formulas in a particular formal logic and signature, while a model is a structure that gives a concrete interpretation of the theory. Model theory is closely related to universal algebra and algebraic geometry, although the methods of model theory focus more on logical considerations than those fields.

The set of all models of a particular theory is called an elementary class; classical model theory seeks to determine the properties of models in a particular elementary class, or determine whether certain classes of structures form elementary classes.

The method of quantifier elimination can be used to show that definable sets in particular theories cannot be too complicated. Tarski (1948) established quantifier elimination for real-closed fields, a result which also shows the theory of the field of real numbers is decidable. (He also noted that his methods were equally applicable to algebraically closed fields of arbitrary characteristic.) A modern subfield developing from this is concerned with o-minimal structures.

Morley's categoricity theorem, proved by Michael D. Morley (1965), states that if a first-order theory in a countable language is categorical in some uncountable cardinality, i.e. all models of this cardinality are isomorphic, then it is categorical in all uncountable cardinalities.

A trivial consequence of the continuum hypothesis is that a complete theory with less than continuum many nonisomorphic countable models can have only countably many. Vaught's conjecture, named after Robert Lawson Vaught, says that this is true even independently of the continuum hypothesis. Many special cases of this conjecture have been established.

Recursion Theory

Recursion theory, also called computability theory, studies the properties of computable functions and the Turing degrees, which divide the uncomputable functions into sets which have the same level of uncomputability. Recursion theory also includes the study of generalized computability and definability. Recursion theory grew from of the work of Alonzo Church and Alan Turing in the 1930s, which was greatly extended by Kleene and Post in the 1940s.

Classical recursion theory focuses on the computability of functions from the natural numbers to the natural numbers. The fundamental results establish a robust, canonical class of computable functions with numerous independent, equivalent characterizations using Turing machines, λ calculus, and other systems. More advanced results concern the structure of the Turing degrees and the lattice of recursively enumerable sets.

Generalized recursion theory extends the ideas of recursion theory to computations that are no longer necessarily finite. It includes the study of computability in higher types as well as areas such as hyperarithmetical theory and á-recursion theory.

Contemporary research in recursion theory includes the study of applications such as algorithmic randomness and computable model theory as well as new results in pure recursion theory.

Algorithmically Unsolvable Problems

An important subfield of recursion theory studies algorithmic unsolvability; a problem is algorithmically unsolvable if there is no computable function which, given any [code for an] instance of the problem, returns the correct answer.

The first results about unsolvability, obtained independently by Church and Turing in 1936, showed that the Entscheidungsproblem is algorithmically unsolvable. Turing proved this by establishing the unsolvability of the halting problem, a result with far-ranging implications in both recursion theory and computer science.

There are many known examples of undecidable problems from ordinary mathematics. The word problem for groups was proved algorithmically unsolvable by Pyotr Sergeyevich Novikov in 1955 and independently by W. Boone in 1959. The busy beaver problem, developed by Tibor Radó in 1962, is another well-known example.

Hilbert's tenth problem asked for an algorithm to determine whether a multivariate polynomial equation with integer coefficients has a solution in the integers. Partial progress was made by Julia Robinson, Martin Davis, and Hilary Putnam.

The algorithmic unsolvability of the problem was proved by Yuri Matiyasevich in 1970.

Proof Theory and Constructive Mathematics

Proof theory is the study of formal proofs in various logical deduction systems. These proofs are represented as formal mathematical objects, facilitating their analysis by mathematical techniques. Several deduction systems are commonly considered, including Hilbert-style deduction systems, systems of natural deduction, and the sequent calculus developed by Gentzen.

The study of constructive mathematics, in the context of mathematical logic, includes the study of systems in non-classical logic such as intuitionistic logic, as well as the study of predicative systems. An early proponent of predicativism was Hermann Weyl, who showed it is possible to develop a large part of real analysis using only predicative methods.

Because proofs are entirely finitary, whereas truth in a structure is not, it is common for work in constructive mathematics to emphasize provability. The relationship between provability in classical (or nonconstructive) systems and provability in intuitionistic (or constructive, respectively) systems is of particular interest. Results such as the Gödel–Gentzen negative translation show that it is possible to embed (or *translate*) classical logic into intuitionistic logic, allowing some properties about intuitionistic proofs to be transferred back to classical proofs.

Recent developments in proof theory include the study of proof mining by Ulrich Kohlenbach and the study of proof-theoretic ordinals by Michael Rathjen.

Connections With Computer Science

The study of computability theory in computer science is closely related to the study of computability in mathematical logic. There is a difference of emphasis, however. Computer scientists often focus on concrete programming languages and feasible computability, while researchers in mathematical logic often focus on computability as a theoretical concept and on noncomputability.

The study of programming language semantics is related to model theory, as is program verification (in particular, model checking). The Curry-Howard isomorphism between proofs and programs relates to proof theory, especially intuitionistic logic. Formal calculi such as the lambda calculus and combinatory logic are now studied as idealized programming languages.

Computer science also contributes to mathematics by developing techniques for the automatic checking or even finding of proofs, such as automated theorem proving and logic programming.

Foundations of Mathematics

In the 19th century, mathematicians became aware of logical gaps and inconsistencies in their field. It was shown that Euclid's axioms for geometry, which had been taught for centuries as an example of the axiomatic method, were incomplete. The use of infinitesimals, and the very definition of function, came into question in analysis, as pathological examples such as Weierstrass' nowhere-differentiable continuous function were discovered.

Cantor's study of arbitrary infinite sets also drew criticism. Leopold Kronecker famously stated "God made the integers; all else is the work of man," endorsing a return to the study of finite, concrete objects in mathematics. Although Kronecker's argument was carried forward by constructivists in the 20th century, the mathematical community as a whole rejected them. David Hilbert argued in favour of the study of the infinite, saying "No one shall expel us from the Paradise that Cantor has created."

Mathematicians began to search for axiom systems that could be used to formalize large parts of mathematics. In addition to removing ambiguity from previously-naive terms such as function, it was hoped that this axiomatization would allow for consistency proofs.

In the 19th century, the main method of proving the consistency of a set of axioms was to provide a model for it. Thus, for example, non-Euclidean geometry can be proved consistent by defining *point* to mean a point on a fixed sphere and *line* to mean a great circle on the sphere. The resulting structure, a model of elliptic geometry, satisfies the axioms of plane geometry except the parallel postulate.

With the development of formal logic, Hilbert asked whether it would be possible to prove that an axiom system is consistent by analysing the structure of possible proofs in the system, and showing through this analysis that it is impossible to prove a contradiction. This idea led to the study of proof theory. Moreover, Hilbert proposed that the analysis should be entirely concrete, using the term *finitary* to refer to the methods he would allow but not precisely defining them.

This project, known as Hilbert's program, was seriously affected by Gödel's incompleteness theorems, which show that the consistency of formal theories of arithmetic cannot be established using methods formalizable in those theories.

Gentzen showed that it is possible to produce a proof of the consistency of arithmetic in a finitary system augmented with axioms of transfinite induction, and the techniques he developed to so do were seminal in proof theory.

A second thread in the history of foundations of mathematics involves nonclassical logics and constructive mathematics. The study of constructive mathematics includes many different programs with various definitions of *constructive*. At the most accommodating end, proofs in ZF set theory that do not use the axiom of choice are called constructive by many mathematicians. More limited versions of constructivism limit themselves to natural numbers, number-theoretic functions, and sets of natural numbers (which can be used to represent real numbers, facilitating the study of mathematical analysis). A common idea is that a concrete means of computing the values of the function must be known before the function itself can be said to exist.

In the early 20th century, Luitzen Egbertus Jan Brouwer founded intuitionism as a philosophy of mathematics. This philosophy, poorly understood at first, stated that in order for a mathematical statement to be true to a mathematician, that person must be able to *intuit* the statement, to not only believe its truth but understand the reason for its truth.

A consequence of this definition of truth was the rejection of the law of the excluded middle, for there are statements that, according to Brouwer, could not be claimed to be true while their negations also could not be claimed true. Brouwer's philosophy was influential, and the cause of bitter disputes among prominent mathematicians. Later, Kleene and Kreisel would study formalized versions of intuitionistic logic (Brouwer rejected formalization, and presented his work in unformalized natural language). With the advent of the BHK interpretation and Kripke models, intuitionism became easier to reconcile with classical mathematics.

Algebra of Sets

The algebra of sets develops and describes the basic properties and laws of sets, the set-theoretic operations of union, intersection, and complementation and the relations of set equality (mathematics) and set inclusion. It also provides systematic procedures for evaluating expressions, and performing calculations, involving these operations and relations.

The algebra of sets is the development of the fundamental properties of set operations and set relations. These properties provide insight into the fundamental nature of sets. They also have practical considerations.

Just like expressions and calculations in ordinary arithmetic, expressions and calculations involving sets can be quite complex. It is helpful to have systematic procedures available for manipulating and evaluating such expressions and performing such computations.

In the case of arithmetic, it is elementary algebra that develops the fundamental properties of arithmetic operations and relations.

For example, the operations of addition and multiplication obey familiar laws such as associativity, commutativity and distributivity, while, the "less than or equal" relation satisfies such laws as reflexivity, antisymmetry and transitivity.

These laws provide tools which facilitate computation, as well as describe the fundamental nature of numbers, their operations and relations.

The algebra of sets is the set-theoretic analogue of the algebra of numbers. It is the algebra of the set-theoretic operations of union, intersection and complementation, and the relations of equality and inclusion.

These are the topics covered in this article. For a basic introduction to sets see the article on sets, for a fuller account see naive set theory, and for a full rigorous axiomatic treatment.

The Fundamental Laws of Set Algebra

The binary operations of set union and intersection satisfy many identities. Several of these identities or "laws" have well established names.

Three pairs of laws, are stated, without proof, in the following proposition.

Proposition 1: For any sets *A*, *B*, and *C*, the following identities hold:

commutative laws:

- $A \cup B = B \cup A$
- $A \cap B = B \cap A$

associative laws:

- $(A \cup B) \cup C = A \cup (B \cup C)$
- $(A \cap B) \cap C = A \cap (B \cap C)$

distributive laws:

- $A \cup (B \cap C) = (A \cup B) \cap (A \cup C)$
- $A \cap (B \cup C) = (A \cap B) \cup (A \cap C)$

Notice that the analogy between unions and intersections of sets, and addition and multiplication of numbers, is quite striking. Like addition and multiplication, the operations of union and intersection are commutative and associative, and intersection *distributes* over unions. However, unlike addition and multiplication, union also *distributes* over intersection.

The next proposition, states two additional pairs of laws involving three specials sets: the empty set, the universal set and the complement of a set.

Proposition 2: For any subset *A* of universal set U, where Ø is the empty set, the following identities hold:

Identity laws:

- $A \cup \varnothing = A$
- $A \cap U = A$

Complement laws:

- $A \cup A^C = U$
- $A \cap A^C = \varnothing$

The identity laws (together with the commutative laws) say that, just like 0 and 1 for addition and multiplication, Ø and U are the identity elements for union and intersection, respectively.

Unlike addition and multiplication, union and intersection do not have inverse elements. However the complement laws give the fundamental properties of the somewhat inverse-like unary operation of set complementation.

The preceding five pairs of laws: the commutative, associative, distributive, identity and complement laws, can be said to encompass all of set algebra, in the sense that every valid proposition in the algebra of sets can be derived from them.

The Principle of Duality

The above propositions display the following interesting pattern. Each of the identities stated above is one of a pair of identities, such that, each can be transformed into the other by interchanging *" and)", and also Ø and U.

These are examples of an extremely important and powerful property of set algebra, namely, the principle of duality for sets, which asserts that for any true statement about sets, the dual statement obtained by interchanging unions and intersections, interchanging U and Ø and reversing inclusions is also true.

A statement is said to be self-dual if it is equal to its own dual.

Some Additional Laws for Unions and Intersections

The following proposition states six more important laws of set algebra, involving unions and intersections.

Proposition 3: For any subsets A and B of a universal set U, the following identities hold:

idempotent laws:

- $A \cup A = A$
- $A \cap A = A$

domination laws:

- $A \cup U = U$
- $A \cap \varnothing = \varnothing$

absorption laws:

- $A \cup (A \cap B) = A$
- $A \cap (A \cup B) = A$

As noted above each of the laws stated in proposition 3, can be derived from the five fundamental pairs of laws stated in proposition 1 and proposition 2. As an illustration, a proof is given below for the idempotent law for union.

Proof:

$A \cup A \quad = (A \cup A) \cap U$ by the identity law of intersection

$= (A \cup A) \cap (A \cup A^C)$ by the complement law for union

$= A \cup (A \cap A^C)$ by the distributive law of union over intersection

$= A \cup \varnothing$ by the complement law for intersection

$= A$ by the identity law for union

The following proof illustrates that the dual of the above proof is the proof of the dual of the idempotent law for union, namely the idempotent law for intersection.

Proof:

$A \cap A \quad = (A \cap A) \cup \varnothing$ by the identity law for union

$= (A \cap A) \cup (A \cap A^C)$ by the complement law for intersection

$= A \cap (A \cap A^C)$ by the distributive law of intersection over union

$= A \cap U$ by the complement law for union

$= A$ by the identity law for intersection

Some Additional Laws for Complements

The following proposition states five more important laws of set algebra, involving complements.

Proposition 4: Let A and B be subsets of a universe U, then:

De Morgan's laws:

- $(A \cup B)^C = A^C \cap B^C$
- $(A \cap B)^C = A^C \cup B^C$

double complement or Involution law:

- $A^{CC} = A$

complement laws for the universal set and the empty set:

- $\varnothing^C = U$
- $U^C = \varnothing$

Notice that the double complement law is self-dual.

The next proposition, which is also self-dual, says that the complement of a set is the only set that satisfies the complement laws. In other words, complementation is characterized by the complement laws.

Proposition 5: Let A and B be subsets of a universe U, then:

uniqueness of complements:

- If $A \cup B = U$, and $A \cap B = \varnothing$, then $B = A^C$

The Algebra of Inclusion

The following proposition says that inclusion is a partial order.

Proposition 6: If *A*, *B* and *C* are sets then the following hold:

reflexivity:

- $A \subseteq A$

antisymmetry:

- $A \subseteq B$ and $B \subseteq A$ if and only if $A = B$

transitivity:

- If $A \subseteq B$ and $B \subseteq C$, then $A \subseteq C$

The following proposition says that for any set *S*, the power set of *S*, ordered by inclusion, is a bounded lattice, and hence together with the distributive and complement laws above, show that it is a Boolean algebra.

Proposition 7: If *A*, *B* and *C* are subsets of a set *S* then the following hold:

existence of a least element and a greatest element:

- $\varnothing \subseteq A \subseteq S$

existence of joins:

- $A \subseteq A \cup B$
- If $A \subseteq C$ and $B \subseteq C$, then $A \cup B \subseteq C$

existence of meets:

- $A \cap B \subseteq A$
- If $C \subseteq A$ and $C \subseteq B$, then $C \subseteq A \cap B$

The following proposition says that the statement $A \subseteq B$ is equivalent to various other statements involving unions, intersections and complements.

Proposition 8: For any two sets *A* and *B*, the following are equivalent:

- $A \subseteq B$
- $A \cap B = A$

- $A \cup B = B$
- $A \setminus B = \varnothing$
- $B^C \subseteq A^C$

The above proposition shows that the relation of set inclusion can be characterized by either of the operations of set union or set intersection, which means that the notion of set inclusion is axiomatically superfluous.

3

Line Geometry

In Euclidean geometry, a line is a straight curve. When geometry is used to model the real world, lines are used to represent straight objects with negligible width and height. Lines are an idealisation of such objects and have no width or height at all and are usually considered to be infinitely long. Lines are a fundamental concept in some approaches to geometry such as Euclid's, but in others such as analytic geometry and Tarski's axioms they enter as derived notions defined in terms of more fundamental primitives such as points.

A *line segment* is a part of a line that is bounded by two distinct end points and contains every point on the line between its end points. Depending on how the line segment is defined, either of the two end points may or may not be part of the line segment. Two or more line segments may have some of the same relationships as lines, such as being parallel, intersecting, or skew.

Euclidean Geometry

When geometry was first formalised by Euclid in *Elements*, he defined lines to be "breadthless length" with a straight line being a line "which lies evenly with the points on itself". These definitions serve little purpose since they use terms which are not, themselves, defined. In fact, Euclid did not use these definitions in work and probably included them just to make it clear to the reader what was being discussed. In modern geometry, a line is simply taken as an undefined object with properties given by postulates.

In an axiomatic formulation of Euclidean geometry, such as that of Hilbert (Euclid's original axioms contained various flaws which have been corrected by modern mathematicians), a line is stated to have certain properties which relate it to other lines and points. For example, for any two distinct points, there is a unique line containing them, and any two distinct lines intersect at most one point. In two dimensions, ie. the Euclidean plane, two lines which do not intersect are called parallel. In higher dimensions, two lines that do not intersect may be parallel if they are contained in a plane, or skew if they are not.

Any collection of lines partitions the plane into convex polygons; this partition is known as an arrangement of lines.

Ray

If you define the concept of "order" of points of a line, you can define a *ray*, or half-line. A ray is part of a line which is finite in one direction, but infinite in the other. It can be defined by two points, the initial point, A, and one other, B. The ray is all the points in the line segment between A and B together with all points, C, on the line through A and B such that the points appear on the line in the order A, B, C.

In topology, a ray in a space X is a continuous embedding $R^+ \rightarrow X$. It is used to define the important concept of end of the space.

Coordinate Geometry

In coordinate geometry, lines in a Cartesian plane can be described algebraically by linear equations and linear functions. In two dimensions, the characteristic equation is often given by the *slope-intercept form*:

$$y = mx + b$$

where:

m is the slope of the line.

b is the y-intercept of the line.

x is the independent variable of the function y.

In three dimensions, a line is described by parametric equations:

$$x = x_0 + at$$

$$y = y_0 + bt$$

$$z = z_0 + ct$$

where:

x, y, and z are all functions of the independent variable t.

x_0, y_0, and z_0 are the initial values of each respective variable.

a, b, and c are related to the slope of the line, such that the vector (a, b, c) is a parallel to the line.

In R^2, every line L is described by a linear equation of the form

$$L = \{(x, y) \mid ax + by = c\}$$

with fixed real coefficients a, b and c such that a and b are not both zero. Important properties of these lines are their slope, x-intercept and y-intercept. The eccentricity of a straight line is infinity.

Euclidean Space

In Euclidean space R^n (and analogously in all other vector spaces), we define a line L as a subset of the form

$$L = \{a + tb \mid t \in \mathbb{R}\}$$

where a and b are given vectors in R^n with b non-zero. The vector b describes the direction of the line, and a is a point on the line. Different choices of a and b can yield the same line.

Projective Geometry

Main article: Projective geometry

In projective geometry, a line is similar to that in Euclidean geometry but has slightly different properties. In many models of projective geometry, the idea of the line rarely conforms to the notion of the "straight curve" as it is visualised in Euclidean geometry. The Poincaré half-plane model is a typical example of this.

Geodesics

The "straightness" of a line, interpreted as the property that it minimizes distances between its points, can be generalized and leads to the concept of geodesics on differentiable manifolds.

Plane (Geometry)

In mathematics, a plane is a flat surface Chyea. Planes can arise as subspaces of some higher dimensional space, as with the walls of a room, or they may enjoy an independent existence in their own right, as in the setting of Euclidean geometry.

When working in two-dimensional Euclidean space, the definite article is used, the plane, to refer to the whole space. Many fundamental tasks in geometry, trigonometry, and graphing are performed in two-dimensional space, or in other words, in the plane. A lot of mathematics can be and has been performed in the plane, notably in the areas of geometry, trigonometry, graph theory and graphing. All two-dimensional figures are assumed to be on a plane, even on *the* plane, unless otherwise specified.

Euclidean Geometry

Euclid set forth the first known axiomatic treatment of geometry. He selected a small core of undefined terms (called *common notions*) and postulates (or axioms) which he then uses these to prove various geometrical statements. Although the plane in its modern sense is not directly given a definition anywhere in the *Elements*, it may be thought of as part of the common notions. In his work Euclid never makes use of numbers to measure length, angle, or area. In this way the Euclidean plane is not quite the same as the Cartesian plane.

Planes in Various Areas of Mathematics

In addition to its familiar geometric structure, with isomorphisms that are isometries with respect to the usual inner product, the plane may be viewed at various other levels of abstraction. Each level of abstraction corresponds to a specific category.

At one extreme, all geometrical and metric concepts may be dropped to leave the topological plane, which may be thought of as an idealized homotopically trivial infinite rubber sheet, which retains a notion of proximity, but has no distances. The topological plane has a concept of a linear path, but no concept of a straight line. The topological plane, or its equivalent the open disc, is the basic topological neighbourhood used to construct surfaces (or 2-manifolds) classified in low-dimensional topology. Isomorphisms of the topological plane are all continuous bijections. The topological plane is the natural context for the branch of graph theory that deals with planar graphs, and results such as the four colour theorem.

The plane may also be viewed as an affine space, whose isomorphisms are combinations of translations and non-singular linear maps. From this viewpoint there are no distances, but colinearity and ratios of distances on any line are preserved.

Differential geometry views a plane as a 2-dimensional real manifold, a topological plane which is provided with a differential structure. Again in this case, there is no notion of distance, but there is now a concept of smoothness of maps, for example a differentiable or smooth path (depending on the type of differential structure applied). The isomorphisms in this case are bijections with the chosen degree of differentiability.

In the opposite direction of abstraction, we may apply a compatible field structure to the geometric plane, giving rise to the complex plane and the major area of complex analysis. The complex field has only two isomorphisms that leave the real line fixed, the identity and conjugation. In the same way as in the real case, the plane may also be viewed as the simplest, one-dimensional (over the complex numbers) complex manifold, sometimes called the complex line. However, this viewpoint contrasts sharply with the case of the plane as a 2-dimensional real manifold. The isomorphisms are all conformal bijections of the complex plane, but the only possibilities are maps that correspond to the composition of a multiplication by a complex number and a translation.

In addition, the Euclidean geometry (which has zero curvature everywhere) is not the only geometry that the plane may have. The plane may be given a spherical geometry by using the stereographic projection. This can be thought of as placing a sphere on the plane (just like a ball on the floor), removing the top point, and projecting the sphere onto the plane from this point). This is one of the projections that may be used in making a flat map of part of the Earth's surface. The resulting geometry has constant positive curvature.

Alternatively, the plane can also be given a metric which gives it constant negative curvature giving the hyperbolic plane. The latter possibility finds an application in the theory of special relativity in the simplified case where there are two spatial dimensions and one time dimension. (The hyperbolic plane is a timelike hypersurface in three-dimensional Minkowski space.)

Topological and Differential Geometric Notions

The one-point compactification of the plane is homeomorphic to a sphere; the open disk is homeomorphic to a sphere with the "north pole" missing; adding that point completes the (compact) sphere. The result of this compactification is a manifold referred as the Riemann sphere or the complex projective line. The projection from the Euclidean plane to a sphere without a point is a diffeomorphism and even a conformal map.

The plane itself is homeomorphic (and diffeomorphic) to an open disk. For the Lobachevsky plane such diffeomorphism is conformal, but for the Euclidean plane it is not.

Planes in Fiction

The 1884 novel Flatland by Edwin A. Abbott features the concept of a geometric, two-dimensional infinite plane inhabited by living geometric figures (triangles, squares, circles, etc.). It has been described by Isaac Asimov, in his foreword to the Signet Classics 1984 edition, as "the best introduction one can find into the manner of perceiving dimensions." this is not a actual plane in the mathematics book

Geometry

The word *geometry* is Greek for *geos* - meaning earth and *metron* - meaning measure. Geometry was extremely important to ancient societies and was used for surveying, astronomy, navigation, and building. Geometry, as we know it is actually known as Euclidean geometry which was written well over 2000 years ago in Ancient Greece by Euclid, Pythagoras, Thales, Plato and Aristotle just to mention a few.

The most fascinating and accurate geometry text was written by Euclid, and was called Elements. Euclid's text has been used for over 2000 years!

Geometry is the study of angles and triangles, perimeter, area and volume. It differs from algebra in that one develops a logical structure where mathematical relationships are proved and applied. In part 1, you will learn about the basic terms associated with Geometry.

Types of Angles

An angle is defined as: where two rays or two line segments join at a common endpoint called the vertex.

Acute Angle

An acute angle measures less that 90° and can look something like these (angles between the grey rays):

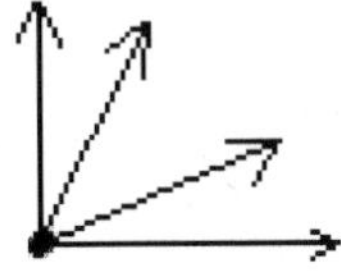

Right Angle

A right angle measures exactly 90° and will look something like this:

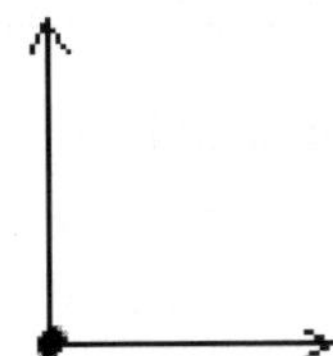

Obtuse Angle

An obtuse angle measures more than 90° but less than 180° and will look something like these:

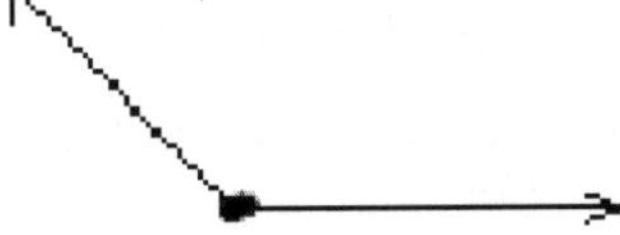

Reflex Angle

An reflex angle is more than 180° but less than 360° and will look something like this:

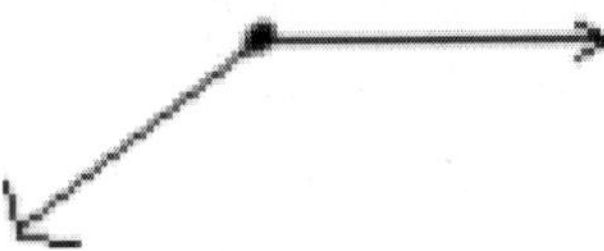

Pairs of Angles:

Complementary Angles

Two angles adding up to 90° are called complementary angles.

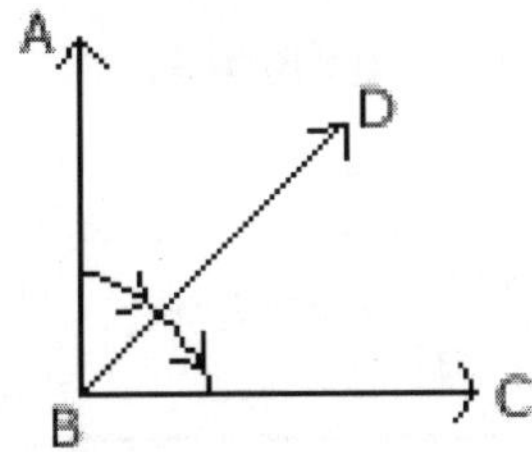

∠ABD + ∠DBC are Complementary

Supplementary Angles

Two angles adding up to 180° are called supplementary angles.

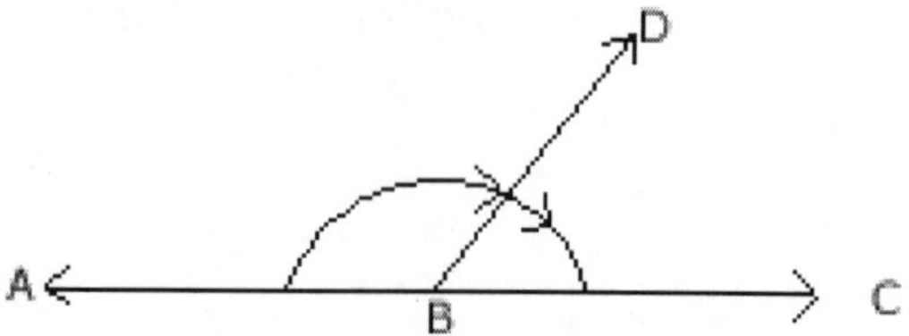

∠ABD + ∠DBC are Supplementary

If you know the angle of ABD, you can easily determine what the DBC is by subtracting ∠ABD from 180!

Some Basics and Important Postulates

Background: Euclid of Alexandria wrote 13 books called 'The Elements' around 300 BC.

These books laid the foundation of geometry. Some of the postulates below were actually posed by Euclid in his 13 books.

They were assumed as axioms, without proof. Euclid's postulates have been slightly corrected over a period of time.

Some are listed here and continue to be part of 'Euclidean Geometry'.

Know this stuff! Learn it, memorize it and keep this page as a handy reference if you expect to understand Geometry.

There are some basic facts, information and postulates that are very important to know in geometry.

Not everything is proved in Geometry, thus we use some *postulates* which are basic assumptions or unproved general statements that we accept.

Here are a few of the basics and postulates that are intended for entry level Geometry.

The Basics and Important Postulates :

1. You can only draw one line between two points. You will not be able to draw a second line through points A and B:

2. There's 360° around a circle.

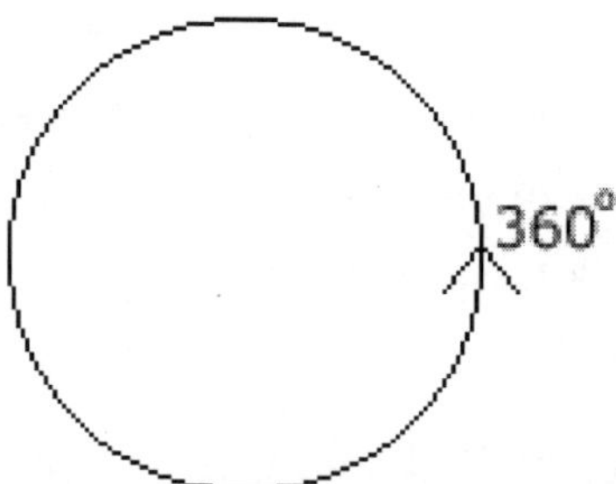

3. Two lines can instersect at ONLY one point. S is the only intersection of and in the figure below.

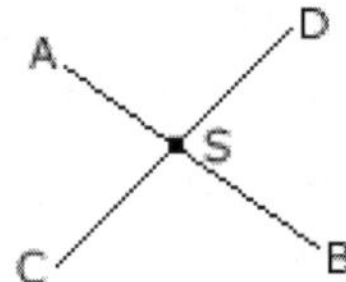

4. A line segment has ONLY one midpoint. M is the only Midpoint of in the figure below.

5. An angle can only have one bisector. (Bisector: a ray that's in the interior of an angle and forms two equal angles with the sides of that angle.)

In the figure below, $\overrightarrow{AD}$ is the bisector of ∠A

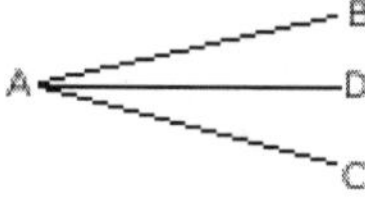

6. Any geometric shape can be moved without changing its shape.

7. A line segment will always be the shortest distance between two points on a plane. The curved line and the bropken line segments are further in distance between A and B.

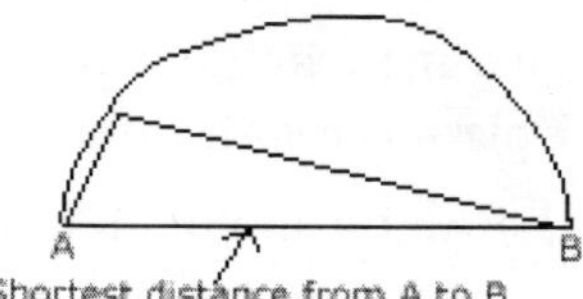

8. If two points lie in a plane, the line containing the points lie in the plane.

9. When two planes intersect, their intersection is a line.

10. ALL lines and planes are sets of points.

11. Every line has a coordinate system. (The Ruler Postulate)

Measuring Angles

The size of an angle will depend on the opening between the two sides of the angle (Pac Mac's mouth) and is measured in units that are referred to as *degrees* which is indicated by the ° symbol.

To help you remember approximate sizes of angles, you will want to remember that a circle, once around measures 360°.

To assist you to remember approximations of angles, it will be helpful to remember the following:

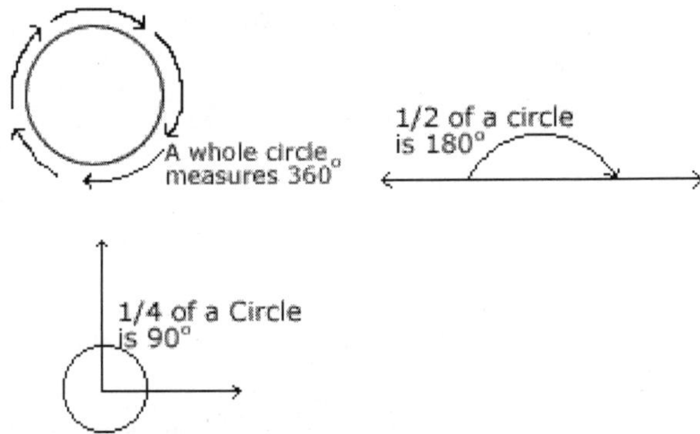

Think of a whole pie as 360°, if you eat a quarter (1/4) of it the measure would be 90°. If you ate 1/2 of the pie? Well, as stated above, 180° is half, or you can add 90° and 90° - the two pieces you ate!

If you cut the whole pie into 8 equal pieces. What angle would one piece of the pie make?

To answer this question, you can divide 360° by 8 *(the total by the number of pieces)*. This will tell you that each piece of pie has a measure of 45°.

Usually, when measuring an angle, you will use a protractor, each unit of measure on a protractor is a degree °.

Note: The size of the angle is not dependent on the lengths of the sides of the angle.

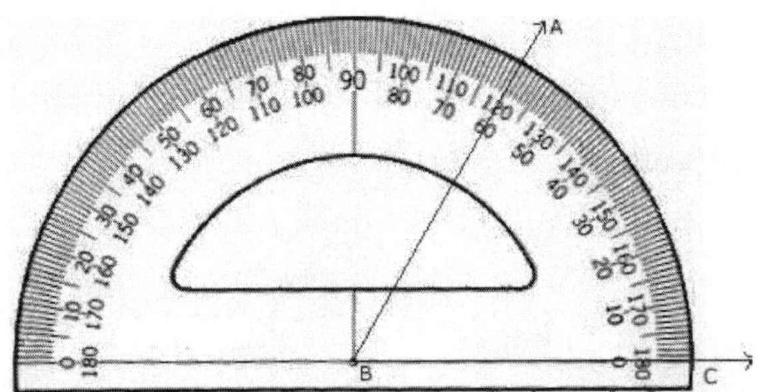

In the above example, the protractor is used to show you that the measure of angle ABC is 60°

Try a few best guesses, the angles below are approximately 10°, 50°, 150°,

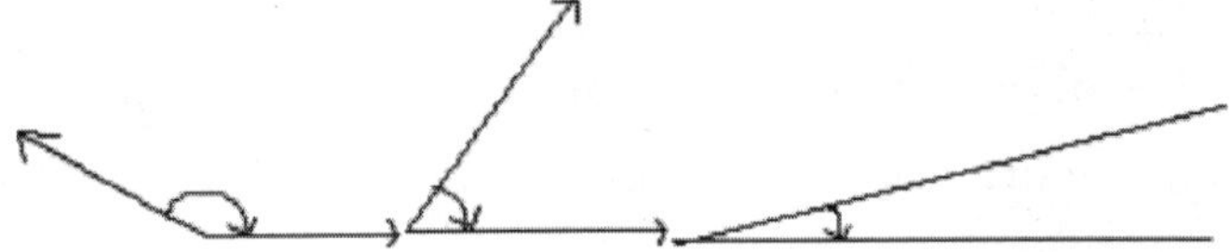

Answers:

1. = approximately 150°

2. = approximately 50°

3 = approximately 10°

More About Angles

1. Congruent angles are angles that have the same number of degrees. For instance, 2 line segments are congurent if they are the same in length. If two angles have the same measure, they too are considered congruent. Symbolically, this can be shown by: as

$\overline{AB} \cong \overline{OP}$ means that Line Segment AB is congruent to line segment OP

2. Bisectors refers to the line, ray or line segment that passes through the midpoint. The bisector divides a segment into two congruent setments as demonstrated here:

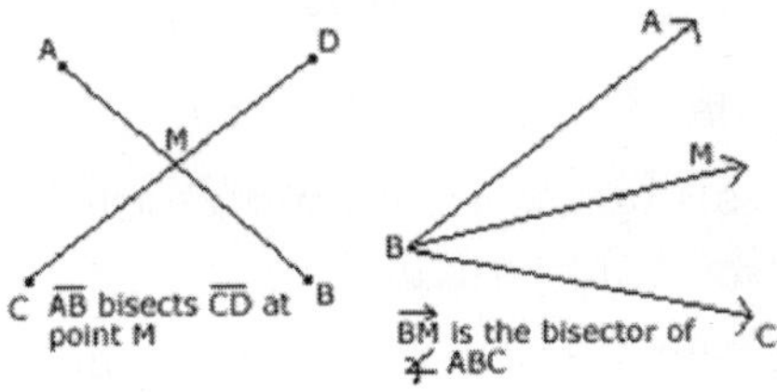

A ray that is in the interior of an angle and divides the original angle into two congruent angles is the bisector of that angle.

3. Exterior and Interior Angles

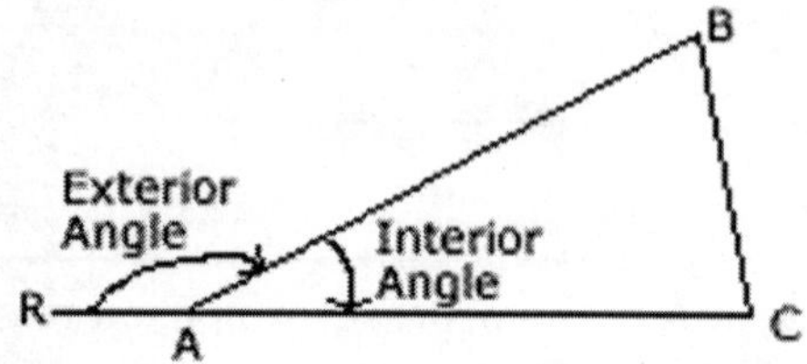

4. Transversal

A Transversal is a line that crosses two parallel lines. In the figure below, A and B are parallel lines.

Note the following when a transversal cuts two parallel lines:

*the four acute angles will be equal

*the four obtuse angles will also be equal

*each acute angle is *supplementary* to eacy obtuse angle

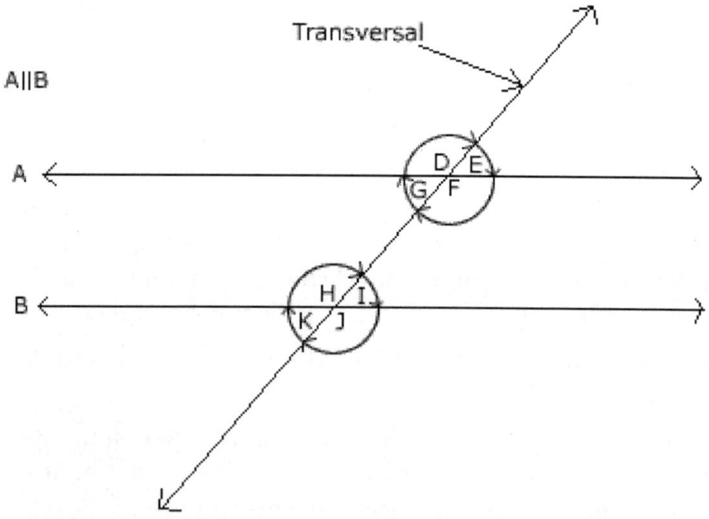

Theorems to Know - *Especially for your SATs!*

1. The sum of the measures of triangles always equals 180°.

You can prove this by using your protractor to measure the three angles, then total the three angles.

See triangle below - 90° + 45° + 45° = 180

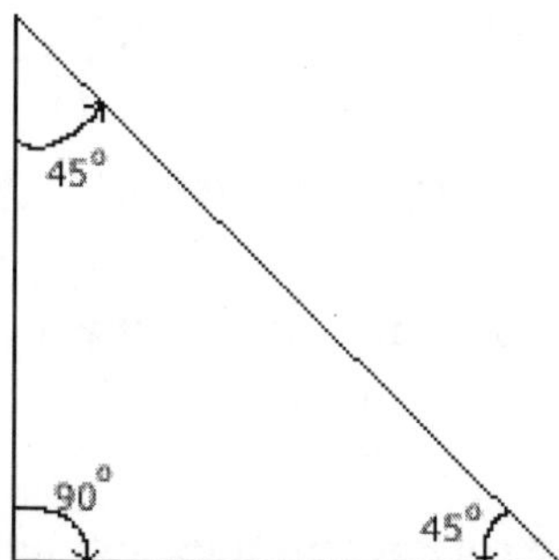

2. The measure of the exterior angle will always equal the sum of the measure of the 2 *remote* interior angles. NOTE: the remote angles in the figure below are b and c.

Therefore, the measure of RAB will be equal to the sum of B and C. If you know the measures B and C then you automatically know what RAB is!

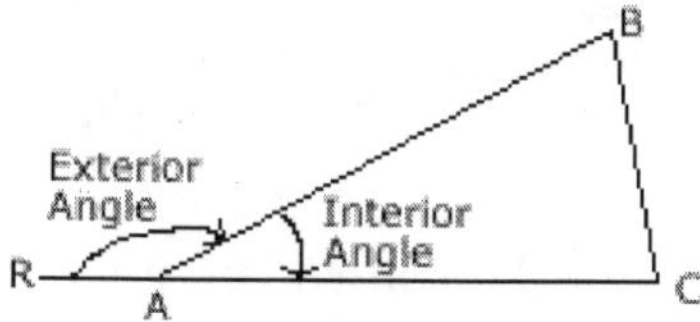

3. If a transversal intersects two lines such that corresponding angles are congruent, then the lines are parallel. AND, If two lines are intersected by a transversal such that interior angles on the same side of the transversal are supplementary, then the lines are parallel.

Non-Euclidean Geometry

In about 300 BC Euclid wrote *The Elements,* a book which was to become one of the most famous books ever written. Euclid stated five postulates on which he based all his theorems:

1. *To draw a straight line from any point to any other.*
2. *To produce a finite straight line continuously in a straight line.*
3. *To describe a circle with any centre and distance.*
4. *That all right angles are equal to each other.*
5. *That, if a straight line falling on two straight lines make the interior angles on the same side less than two right angles, if produced indefinitely, meet on that side on which are the angles less than the two right angles.*

It is clear that the fifth postulate is different from the other four. It did not satisfy Euclid and he tried to avoid its use as long as possible - in fact the first 28 propositions of *The Elements* are proved without using it.

Another comment worth making at this point is that Euclid, and many that were to follow him, assumed that straight lines were infinite.

Proclus (410-485) wrote a commentary on *The Elements* where he comments on attempted proofs to deduce the fifth postulate from the other four, in particular he notes that Ptolemy had produced a false 'proof'.

Proclus then goes on to give a false proof of his own. However he did give the following postulate which is equivalent to the fifth postulate.

Playfair's Axiom:- *Given a line and a point not on the line, it is possible to draw exactly one line through the given point parallel to the line.*

Although known from the time of Proclus, this became known as *Playfair's Axiom* after John Playfair wrote a famous commentary on Euclid in 1795 in which he proposed replacing Euclid's fifth postulate by this axiom.

Many attempts were made to prove the fifth postulate from the other four, many of them being accepted as proofs for long periods of time until the mistake was found. Invariably the mistake was assuming some 'obvious' property which turned out to be equivalent to the fifth postulate.

One such 'proof' was given by Wallis in 1663 when he thought he had deduced the fifth postulate, but he had actually shown it to be equivalent to:-

To each triangle, there exists a similar triangle of arbitrary magnitude.

One of the attempted proofs turned out to be more important than most others.

It was produced in 1697 by Girolamo Saccheri. The importance of Saccheri's work was that he assumed the fifth postulate false and attempted to derive a contradiction.

Saccheri's quadrilateral

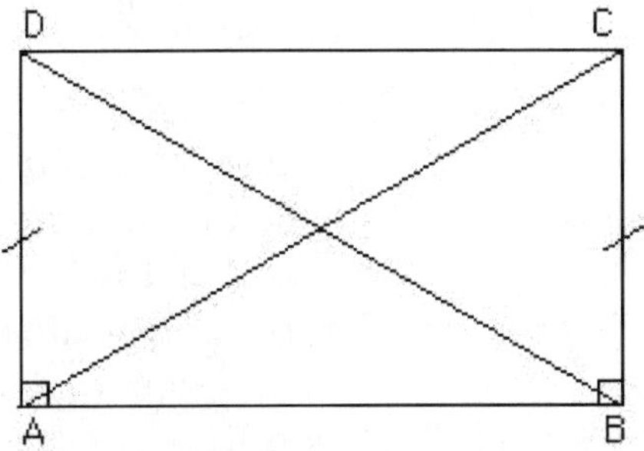

△ABD is congruent to △BAC
(Two sides and included ∠)
Hence AC = BD so △ADC is
congruent to △BCD
(Three sides)
Therefore ∠ADC = ∠BCD.

Figure: *Here is the Saccheri quadrilateral*

In this figure Saccheri proved that the summit angles at *D* and *C* were equal.The proof uses properties of congruent triangles which Euclid proved in Propositions 4 and 8 which are proved before the fifth postulate is used. Saccheri has shown:

a) *The summit angles are* > 90° *(hypothesis of the obtuse angle).*

b) *The summit angles are* < 90° *(hypothesis of the acute angle).*

c) *The summit angles are* = 90° *(hypothesis of the right angle).*

Euclid's fifth postulate is c). Saccheri proved that the hypothesis of the obtuse angle implied the fifth postulate, so obtaining a contradiction. Saccheri then studied the hypothesis of the acute angle and derived many theorems of non-Euclidean geometry without realising what he was doing. However he eventually 'proved' that the hypothesis of the acute angle led to a contradiction by assuming that there is a 'point at infinity' which lies on a plane.

In 1766 Lambert followed a similar line to Saccheri. However he did not fall into the trap that Saccheri fell into and investigated the hypothesis of the acute angle without obtaining a contradiction. Lambert noticed that, in this new geometry, the angle sum of a triangle increased as the area of the triangle decreased.

Legendre spent 40 years of his life working on the parallel postulate and the work appears in appendices to various editions of his highly successful geometry book *Eléments de Géométrie*. Legendre proved that Euclid's fifth postulate is equivalent to:-

The sum of the angles of a triangle is equal to two right angles.

Legendre showed, as Saccheri had over 100 years earlier, that the sum of the angles of a triangle cannot be greater than two right angles. This, again like Saccheri, rested on the fact that straight lines were infinite. In trying to show that the angle sum cannot be less than 180° Legendre assumed that through any point in the interior of an angle it is always possible to draw a line which meets both sides of the angle. This turns out to be another equivalent form of the fifth postulate, but Legendre never realised his error himself.

Elementary geometry was by this time engulfed in the problems of the parallel postulate. D'Alembert, in 1767, called it *the scandal of elementary geometry.*

The first person to really come to understand the problem of the parallels was Gauss. He began work on the fifth postulate in 1792 while only 15 years old, at first attempting to prove the parallels postulate from the other four. By 1813 he had made little progress and wrote:

In the theory of parallels we are even now not further than Euclid. This is a shameful part of mathematics...

However by 1817 Gauss had become convinced that the fifth postulate was independent of the other four postulates. He began to work out the consequences of a geometry in which more than one line can be drawn through a given point parallel to a given line. Perhaps most surprisingly of all Gauss never published this work but kept it a secret. At this time thinking was dominated by Kant who had stated that Euclidean geometry is *the inevitable necessity of thought* and Gauss disliked controversy.

Gauss discussed the theory of parallels with his friend, the mathematician Farkas Bolyai who made several false proofs of the parallel postulate. Farkas Bolyai taught his son, János Bolyai, mathematics but, despite advising his son *not to waste one hour's time on that problem* of the problem of the fifth postulate, János Bolyai did work on the problem.

In 1823 János Bolyai wrote to his father saying *I have discovered things so wonderful that I was astounded... out of nothing I have created a strange new world.* However it took Bolyai a further two years before it was all written down and he published his *strange new world* as a 24 page appendix to his father's book, although just to confuse future generations the appendix was published before the book itself.

Gauss, after reading the 24 pages, described János Bolyai in these words while writing to a friend: *I regard this young geometer Bolyai as a genius of the first order.* However in some sense Bolyai only assumed that the new geometry was possible. He then followed the consequences in a not too dissimilar fashion from those who had chosen to assume the fifth postulate was false and seek a contradiction.

However the real breakthrough was the belief that the new geometry was possible. Gauss, however impressed he sounded in the above quote with Bolyai, rather devastated Bolyai by telling him that he (Gauss) had discovered all this earlier but had not published. Although this must undoubtedly have been true, it detracts in no way from Bolyai's incredible breakthrough.

Nor is Bolyai's work diminished because Lobachevsky published a work on non-Euclidean geometry in 1829. Neither Bolyai nor Gauss knew of Lobachevsky's work, mainly because it was only published in Russian in the *Kazan Messenger* a local university publication. Lobachevsky's attempt to reach a wider audience had failed when his paper was rejected by Ostrogradski.

In fact Lobachevsky fared no better than Bolyai in gaining public recognition for his momentous work. He published *Geometrical investigations on the theory of parallels* in 1840 which, in its 61 pages, gives the clearest account of Lobachevsky's work. The publication of an account in French in Crelle's Journal in 1837 brought his work on non-Euclidean geometry to a wide audience but the mathematical community was not ready to accept ideas so revolutionary.

In Lobachevsky's 1840 booklet he explains clearly how his non-Euclidean geometry works.

> *All straight lines which in a plane go out from a point can, with reference to a given straight line in the same plane, be divided into two classes - into cutting and non-cutting. The boundary lines of the one and the other class of those lines will be called parallel to the given line.*

AD is the perpendicular from A to BC.
AE is perpendicular to AD.
Within the angle EAD, some lines (such as AF) will meet BC.
Assume that AE is not the only line which does not meet BC, so let AG be another such line.
AF is a cutting line and AG is a non-cutting line.
There must be a boundary between cutting and non-cutting lines and we may take AH as this boundary.

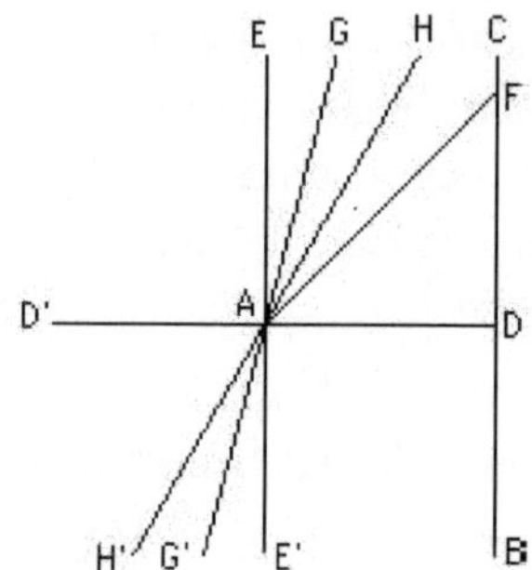

Figure: *Here is the Lobachevsky's diagram*

Hence Lobachevsky has replaced the fifth postulate of Euclid by:-

Lobachevsky's Parallel Postulate. *There exist two lines parallel to a given line through a given point not on the line.*

Lobachevsky went on to develop many trigonometric identities for triangles which held in this geometry, showing that as the triangle became small the identities tended to the usual trigonometric identities.

Riemann, who wrote his doctoral dissertation under Gauss's supervision, gave an inaugural lecture on 10 June 1854 in which he reformulated the whole concept of geometry which he saw as a space with enough extra structure to be able to measure things like length. This lecture was not published until 1868, two years after Riemann's death but was to have a profound influence on the development of a wealth of different geometries. Riemann briefly discussed a 'spherical' geometry in which every line through a point *P* not on a line *AB* meets the line *AB*. In this geometry no parallels are possible.

It is important to realise that neither Bolyai's nor Lobachevsky's description of their new geometry had been proved to be consistent. In fact it was no different from Euclidean geometry in this respect although the many centuries of work with Euclidean geometry was sufficient to convince mathematicians that no contradiction would ever appear within it.

The first person to put the Bolyai - Lobachevsky non-Euclidean geometry on the same footing as Euclidean geometry was Eugenio Beltrami (1835-1900). In 1868 he wrote a paper *Essay on the interpretation of non-Euclidean geometry* which produced a model for 2-dimensional non-Euclidean geometry within 3-dimensional Euclidean geometry. The model was obtained on the surface of revolution of a tractrix about its asymptote. This is sometimes called a *pseudo-sphere.*

You can see the graph of a tractrix and what the top half of a Pseudo-sphere looks like.

In fact Beltrami's model was incomplete but it certainly gave a final decision on the fifth postulate of Euclid since the model provided a setting in which Euclid's first four postulates held but the fifth did not hold. It reduced the problem of consistency of the axioms of non-Euclidean geometry to that of the consistency of the axioms of Euclidean geometry.

Beltrami's work on a model of Bolyai - Lobachevsky's non-Euclidean geometry was completed by Klein in 1871. Klein went further than this and gave models of other non-Euclidean geometries such as Riemann's spherical geometry. Klein's work was based on a notion of distance defined by Cayley in 1859 when he proposed a generalised definition for distance.

Klein showed that there are three basically different types of geometry. In the Bolyai - Lobachevsky type of geometry, straight lines have two infinitely distant points. In the Riemann type of spherical geometry, lines have no (or more precisely two imaginary) infinitely distant points. Euclidean geometry is a limiting case between the two where for each line there are two coincident infinitely distant points.

Conic Section

In mathematics, a conic section (or just conic) is a curve obtained by intersecting a cone (more precisely, a circular conical surface) with a plane. A conic section is therefore a restriction of a quadric surface to the plane. The conic sections were named and studied as long ago as 200 BC, when Apollonius of Perga undertook a systematic study of their properties.

Types

The three types of conics are the hyperbola, ellipse, and parabola. The circle can be considered as a fourth type (as it was by Apollonius) or as a kind of ellipse. The circle and the ellipse arise when the intersection of cone and plane is a closed curve. The circle is obtained when the cutting plane is parallel to the plane of the generating circle of the cone — for a right cone as in the picture at the top of the page this means that the cutting plane is perpendicular to the symmetry axis of the cone. If the cutting plane is parallel to exactly one generating line of the cone, then the conic is unbounded and is called a parabola.

In the remaining case, the figure is a hyperbola. In this case, the plane will intersect *both* halves (*nappes*) of the cone, producing two separate unbounded curves, though often one is ignored.

Degenerate Cases

There are multiple degenerate cases, in which the plane passes through the apex of the cone. The intersection in these cases can be a straight line (when the plane is tangential to the surface of the cone); a point (when the angle between the plane and the axis of the cone is larger than tangential); or a pair of intersecting lines (when the angle is smaller).

Where the cone is a cylinder, i.e. with the vertex at infinity, cylindric sections are obtained. Although these yield mostly ellipses (or circles) as usual, a degenerate case of two parallel lines, known as a ribbon, can also be produced, and it is also possible for there to be no intersection at all.

Eccentricity

The four defining conditions above can be combined into one condition that depends on a fixed point F (the *focus*), a line L (the *directrix*) not containing F and a nonnegative real number e (the *eccentricity*).

The corresponding conic section consists of all points whose distance to F equals e times their distance to L. For $0 < e < 1$ we obtain an ellipse, for $e = 1$ a parabola, and for $e > 1$ a hyperbola.

For an ellipse and a hyperbola, two focus-directrix combinations can be taken, each giving the same full ellipse or hyperbola. The distance from the centre to the directrix is a/e, where is the semi-major axis of the ellipse, or the distance from the centre to the tops of the hyperbola. The distance from the centre to a focus is.

In the case of a circle, the eccentricity $e = 0$, and one can imagine the directrix to be infinitely far removed from the centre. However, the statement that the circle consists of all points whose distance is e times the distance to L is not useful, because we get zero times infinity.

The eccentricity of a conic section is thus a measure of how far it deviates from being circular.

For a given, the closer is to 1, the smaller is the semi-minor axis.

Cartesian Coordinates

In the Cartesian coordinate system, the graph of a quadratic equation in two variables is always a conic section, and all conic sections arise in this way. The equation will be of the form

$Ax^2 + Bxy + Cy^2 + Dx + Ey + F = 0$ with A, B, C not all zero.

then:

- if $B^2 - 4AC < 0$, the equation represents an ellipse (unless the conic is degenerate, for example $x^2 + y^2 + 10 = 0$);
  - o if $A = C$ and $B = 0$, the equation represents a circle;
- if $B^2 - 4AC = 0$, the equation represents a parabola;
- if $B^2 - 4AC > 0$, the equation represents a hyperbola;
  - o if we also have $A + C = 0$, the equation represents a rectangular hyperbola.

Note that A and B are just polynomial coefficients, not the lengths of semi-major/minor axis as defined in the previous sections.

Through change of coordinates these equations can be put in standard forms:

- Circle: $x^2 + y^2 = a^2$
- Ellipse: $\frac{x^2}{a^2} + \frac{y^2}{b^2} = 1$ $\frac{x^2}{b^2} + \frac{y^2}{a^2} = 1$
- Parabola: $y^2 = 4ax$
- Hyperbola: $\frac{x^2}{a^2} - \frac{y^2}{b^2} = 1$
- Rectangular Hyperbola: $xy = c^2$

Such forms will be symmetrical about the x-axis and for the circle, ellipse and hyperbola symmetrical about the y-axis.

The rectangular hyperbola however is only symmetrical about the lines $y = x$ and $y = -x$. Therefore its inverse function is exactly the same as its original function.

These standard forms can be written as parametric equations,

- Circle: $(a\cos\theta, a\sin\theta)$,
- Ellipse: $(a\cos\theta, b\sin\theta)$,
- Parabola: $(at^2, 2at)$,

- Hyperbola: $(a\sec\theta, b\tan\theta)$ or $(\pm a\cosh u, b\sinh u)$.
- Rectangular Hyperbola: $(ct, \frac{c}{t})$

Homogeneous Coordinates

In homogeneous coordinates a conic section can be represented as:

$$A_1x^2 + A_2y^2 + A_3z^2 + 2B_1xy + 2B_2xz + 2B_3yz = 0.$$

Or in matrix notation

$$\begin{bmatrix} x & y & z \end{bmatrix} . \begin{bmatrix} A_1 & B_1 & B_2 \\ B_1 & A_2 & B_3 \\ B_2 & B_3 & A_3 \end{bmatrix} . \begin{bmatrix} x \\ y \\ z \end{bmatrix} = 0.$$

The matrix $M = \begin{bmatrix} A_1 & B_1 & B_2 \\ B_1 & A_2 & B_3 \\ B_2 & B_3 & A_3 \end{bmatrix}$ is called *the matrix of the conic section*.

$\Delta = \det(M) = \det\left(\begin{bmatrix} A_1 & B_1 & B_2 \\ B_1 & A_2 & B_3 \\ B_2 & B_3 & A_3 \end{bmatrix}\right)$ is called the determinant of the conic section. If $\Delta = 0$ then the *conic section* is said to be *degenerate*, this means that the conic section is in fact a union of two straight lines. A conic section that intersects itself is always degenerate, however not all degenerate conic sections intersect themselves, if they do not they are straight lines.

For example, the conic section $\begin{bmatrix} x & y & z \end{bmatrix} . \begin{bmatrix} 1 & 0 & 0 \\ 0 & -1 & 0 \\ 0 & 0 & 0 \end{bmatrix} . \begin{bmatrix} x \\ y \\ z \end{bmatrix} = 0$ reduces to the union of two lines:

$$\{x^2 - y^2 = 0\} = \{(x+y)(x-y) = 0\} = \{x+y=0\} \cup \{x-y=0\}.$$

Similarly, a conic section sometimes reduces to a (single) line:

$$\{x^2 + 2xy + y^2 = 0\} = \{(x+y)^2 = 0\} = \{x+y=0\} \cup \{x+y=0\} = \{x+y=0\}.$$

$\delta = \det\left(\begin{bmatrix} A_1 & B_1 \\ B_1 & A_2 \end{bmatrix}\right)$ is called the discriminant of the conic section.

If $\delta = 0$ then the *conic section* is a parabola, if $\delta<0$, it is an hyperbola and if $\delta>0$, it is an ellipse.

A conic section is a circle if $\delta>0$ and $A_1 = A_2$, it is an rectangular hyperbola if $\delta<0$ and $A_1 = -A_2$.

It can be proven that in the complex projective plane CP^2 two conic sections have four points in common (if one accounts for multiplicity), so there are never more than 4 intersection points and there is always 1 *intersection point* (possibilities: 4 distinct intersection points, 2 singular intersection points and 1 double intersection points, 2 double intersection points, 1 singular intersection point and 1 with multiplicity 3, 1 intersection point with multiplicity 4).

If there exists at least one intersection point with multiplicity > 1, then the two conic sections are said to be tangent.

If there is only one intersection point, which has multiplicity 4, the two conic sections are said to be osculating.

Furthermore each straight line intersects each conic section twice. If the intersection point is double, the line is said to be tangent and it is called the tangent line. Because every straight line intersects a conic section twice, each conic section has two points at infinity (the intersection points with the line at infinity).

If these points are real, the conic section must be a hyperbola, if they are imaginary conjugated, the conic section must be an ellipse, if the conic section has one double point at infinity it is a parabola.

If the points at infinity are (1,i,0) and (1,-i,0), the conic section is a circle. If a conic section has one real and one imaginary point at infinity or it has two imaginary points that are not conjugated it is neither a parabola nor an ellipse nor a hyperbola.

Polar Coordinates

In polar coordinates, a conic section with one focus at the origin and, if any, the other on the *x*-axis, is given by the equation

$$r = \frac{l}{(1 + e\cos\theta)}$$

where e is the eccentricity and l is the semi-latus rectum. As above, for $e = 0$, we have a circle, for $0 < e < 1$ we obtain an ellipse, for $e = 1$ a parabola, and for $e > 1$ a hyperbola.

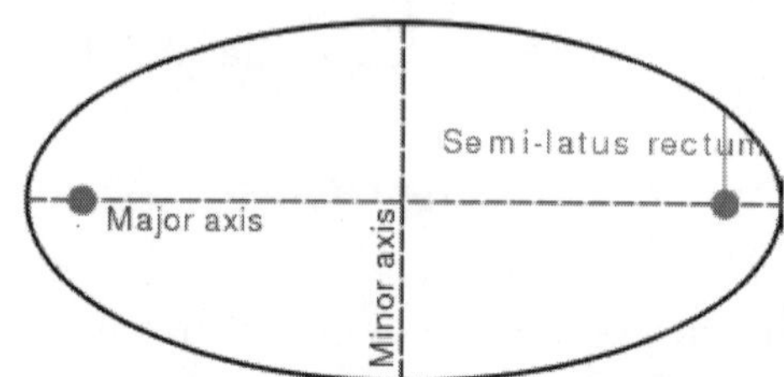

Figure: *Semi-latus rectum in the case of an ellipse*

Properties

Conic sections are always "smooth". More precisely, they never contain any inflection points. This is important for many applications, such as aerodynamics, where a smooth surface is required to ensure laminar flow and to prevent turbulence.

Applications

Conic sections are important in astronomy: the orbits of two massive objects that interact according to Newton's law of universal gravitation are conic sections if their common centre of mass is considered to be at rest. If they are bound together, they will both trace out ellipses; if they are moving apart, they will both follow parabolas or hyperbolas.

In projective geometry, the conic sections in the projective plane are equivalent to each other up to projective transformations.

The solutions to a two second degree equations system in two variables may be seen as the coordinates of the intersections of two generic conic sections. In particular two conics may possess none, two or four possibly coincident intersection points.

The best method of locating these solutions exploits the homogeneous matrix representation of conic sections, i.e. a 3x3 symmetric matrix which depends on six parameters.

The procedure to locate the intersection points follows these steps:

- given the two conics C_1 and C_2 consider the pencil of conics given by their linear combination λC_1 + ìC_2
- identify the homogeneous parameters (λ,ì) which corresponds to the degenerate conic of the pencil. This can be done by imposing that *det*(λC_1 + ìC_2) = 0, which turns out to be the solution to a third degree equation.

- given the degenerate conic C_0, identify the two, possibly coincident, lines constituting it
- intersects each identified line with one of the two original conic; this step can be done efficiently using the dual conic representation of C_0
- the points of intersection will represent the solution to the initial equation system

Dandelin Spheres

See Dandelin spheres for a short elementary argument showing that the characterization of these curves as intersections of a plane with a cone is equivalent to the characterization in terms of foci, or of a focus and a directrix.

Matrix Representation of Conic Sections

In mathematics, the matrix representation of conic sections is one way of studying a conic section, its axis, vertices, foci, tangents, and the relative position of a given point.

We can also study conic sections whose axes aren't parallel to our coordinate system.

Conic sections have the form of a second-degree polynomial:

$$Q \underline{\underline{\text{def}}}\ Ax^2 + Bxy + Cy^2 + Dx + Ey + F = 0.$$

That can be written as:

$$x^T A_Q x = 0$$

Where X is the homogeneous coordinate vector:

$$\begin{pmatrix} x \\ y \\ 1 \end{pmatrix}$$

And A_Q a matrix:

$$A_Q = \begin{pmatrix} A & B/2 & D/2 \\ B/2 & C & E/2 \\ D/2 & E/2 & F \end{pmatrix}$$

Classification

Regular and degenerated conic sections can be distinguished based on the determinant of A_Q.

If $\det A_Q = 0$, the conic is degenerate.

If Q isn't degenerate, we can see what type of conic section it is by computing the minor $\det A_{33}$ (that is, the determinant of the submatrix resulting from removing the last row and the last column of A_Q):

$$A_{33} = \begin{bmatrix} A & B/2 \\ B/2 & C \end{bmatrix}.$$

- If and only if $|A_{33}| < 0$, it is a hyperbola.
- If and only if $|A_{33}| = 0$, it is a parabola.
- If and only if $|A_{33}| > 0$, it is an ellipse.

In the case of an ellipse, we can make a further distinction between an ellipse and a circle by comparing the last two diagonal elements corresponding to x^2 and y^2.

Moreover, in the case of a non-degenerate ellipse (with $\det A_{33} > 0$ and $\det A_Q \neq 0$), we have a real ellipse if $(A+C)\det A_Q < 0$ but an imaginary ellipse if $(A+C)\det A_Q > 0$. An example of the latter is $x^2 + y^2 + 10 = 0$, which has no real-valued solutions.

If the conic section is degenerate ($\det A_Q = 0$), $\det A_{33}$ still allows us to distinguish its form:

- If and only if $\det A_{33} < 0$, it is two intersecting lines.
- If and only if $\det A_{33} = 0$, it is two parallel straight lines. These lines are distinct and real if $D^2 + E^2 > 4(A+C)F$, coincident if $D^2 + E^2 = 4(A+C)F$, and distinct and imaginary if $D^2 + E^2 < 4(A+C)F$.
- If and only if , it is a single point.

Centre

In the center of the conic, the gradient of the quadratic form Q vanishes, so: $\nabla Q = [\frac{\partial Q}{\partial x}, \frac{\partial Q}{\partial y}] = [0,0]$.

We can calculate the center by taking the first two rows of the associated matrix A_Q, multiplying each by $(x, y, 1)^T$, setting both inner products equal to 0, and solving the system.

$$S \stackrel{\text{def}}{=} \begin{cases} a_{11} + a_{12}x + a_{13}y &= 0 \\ a_{21} + a_{22}x + a_{23}y &= 0 \end{cases} \stackrel{\text{def}}{=} \begin{cases} D/2 + Ax + (B/2)y &= 0 \\ E/2 + (B/2)x + Cy &= 0 \end{cases}$$

This becomes

$$\begin{pmatrix} x_c \\ y_c \end{pmatrix} = \begin{pmatrix} A & B/2 \\ B/2 & C \end{pmatrix}^{-1} \begin{pmatrix} -D/2 \\ -E/2 \end{pmatrix} = \begin{pmatrix} (BE - 2CD)/(4AC - B^2) \\ (DB - 2AE)/(4AC - B^2) \end{pmatrix}$$

Note that in the case of a parabola, defined by $(4AC\text{-}B^2) = 0$, there is no center since the above denominators become zero.

Axes

The major and minor axes are two lines determined by the centre of the conic as a point and eigenvectors of the associated matrix as vectors of direction.

$$a_{1,2} \stackrel{\text{def}}{=} \begin{cases} S(x_0, y_0) & \text{(center of the conic)} \\ \vec{u}(u_x, u_y) & \text{(eigenvector of } A_{33}) \end{cases}$$

So we can write a canonical equation:

$$a_{1,2} \stackrel{\text{def}}{=} \frac{x - x_0}{u_x} = \frac{y - y_0}{u_y}$$

Because a 2 × 2 matrix has 2 eigenvectors, we obtain 2 axes.

Vertices

For a general conic we can determine its vertices by calculating the intersection of the conic and its axes — in other words, by solving the system:

$$V \stackrel{\text{def}}{=} \begin{cases} e & \text{(axis)} \\ Q & \text{(the general equation of the conic)} \end{cases}$$

Tangents

Through a given point, *P*, there are generally two lines tangent to a conic. Expressing *P* as a column vector, p, the two points of tangency are the intersections of the conic with the line whose equation is

$$\mathrm{p}^T A_Q \mathrm{x} = 0$$

When P is on the conic, the line is the tangent there. When P is inside an ellipse, the line is the set of all points whose own associated line passes through P. This line is called the *polar* of the *pole P* with respect to the conic.

Just as P uniquely determines its polar line (with respect to a given conic), so each line determines a unique P. This is thus an expression of geometric duality between points and lines in the plane.

As special cases, the centre of a conic is the pole of the line at infinity, and each asymptote of a hyperbola is a polar (a tangent) to one of its points at infinity.

Using the theory of poles and polars, the problem of finding the four mutual tangents of two conics reduces to finding the intersection of two conics.

Reduced Equation

The *reduced equation of a conic section* is the equation of a conic section translated and rotated so that its centre lies in the centre of the coordinate system and its axes are parallel to the coordinate axes. This is equivalent to saying that the coordinates are moved to satisfy these properties. See the figure.

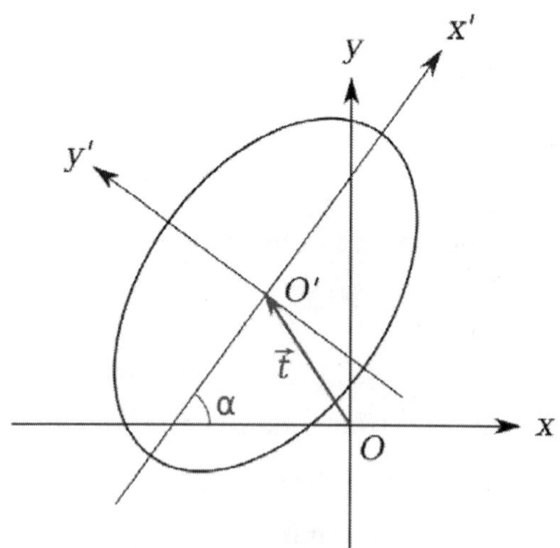

If λ_1 and λ_2 are the eigenvalues of the matrix A_{33}, the reduced equation can be written as:

$$\lambda_1 x'^2 + \lambda_2 y'^2 + \frac{\det A_Q}{\det A_{33}} = 0$$

Dividing by $-\dfrac{\det A_Q}{\det A_{33}}$ we obtain a reduced canonical equation. For example, for an ellipse:

$$\frac{x'^2}{a^2}+\frac{y'^2}{b^2}=1.$$

From here we get 'a' and 'b'.

The transformation of coordinates is given by:

$$T : RS(O,X,Y) \mapsto (O' = S,X',Y') \overset{\text{def}}{=} \begin{cases} \vec{t} & = \overrightarrow{OO'} = S \\ \alpha & = \arccos \dfrac{\vec{a}_1 \cdot \begin{pmatrix} 1 \\ 0 \end{pmatrix}}{|\vec{a}_1|} \end{cases}$$

Quadratic Function and Polynomial

Quadratic Function

A quadratic function, in mathematics, is a polynomial function of the form

$$f(x) = ax^2 + bx + c,\ a \neq 0.$$

The graph of a quadratic function is a parabola whose major axis is parallel to the y-axis.

The expression $ax^2 + bx + c$ in the definition of a quadratic function is a polynomial of degree 2 or second order, or a 2nd degree polynomial, because the highest exponent of x is 2.

If the quadratic function is set equal to zero, then the result is a quadratic equation. The solutions to the equation are called the roots of the equation.

Origin of Word

The adjective *quadratic* comes from the Latin word *quadratum* for square. A term like x^2 is called a square in algebra because it is the area of a *square* with side x.

In general, a prefix quadr(i)- indicates the number 4. Examples are quadrilateral and quadrant.

Quadratum is the Latin word for square because a square has four sides.

Roots

The roots (zeros) of the quadratic function:

$$f(x) = ax^2 + bx + c,$$

are the values of x for which $f(x) = 0$. When the coefficients a, b, and c, are real or complex, then the roots are given by the quadratic formula:

$$\frac{-b \pm \sqrt{b^2 - 4ac}}{2a}$$

Forms of a Auadratic Function

A quadratic function can be expressed in three formats:

- $f(x) = ax^2 + bx + c$ is called the general form or polynomial form,
- $f(x) = a(x - r_1)(x - r_2)$ is called the factored form, where r_1 and r_2 are the roots of the quadratic equation, it is used in logistic map
- $f(x) = a(x - h)^2 + k$ is called the standard form or vertex form where h and k are the x and y coordinates of the vertex.

To convert the general form to factored form, one needs only the quadratic formula to determine the two roots r_1 and r_2. To convert the general form to standard form, one needs a process called completing the square. To convert the factored form (or standard form) to general form, one needs to multiply, expand and/or distribute the factors.

X–intercepts

The x-intercepts of the graph are the same as the roots of the quadratic function.

Vertex

The vertex of a parabola is the place where it turns, hence, it's also called the turning point. If the quadratic function is in standard form, the vertex is (h, k). By the method of completing the square, one can turn the general form:

$f(x) = ax^2 + bx + c$ to

$$f(x) = a\left(x + \frac{b}{2a}\right)^2 - \frac{b^2 - 4ac}{4a},$$

so the vertex of the parabola in the general form will be

$$\left(-\frac{b}{2a}, \frac{\Delta}{4a}\right).$$

If the quadratic function is in factored form

$$f(x) = a(x - r_1)(x - r_2)$$

the average of the two roots, i.e.,

$$\frac{r_1 + r_2}{2}$$

is the *x*-coordinate of the vertex, and hence the vertex is

$$\left(\frac{r_1 + r_2}{2}, f\left(\frac{r_1 + r_2}{2}\right)\right).$$

The vertex is also the maximum point if a < 0 or the minimum point if a > 0..

The Vertical Line

$$x = h = -\frac{b}{2a}$$

that passes through the vertex is also the axis of symmetry of the parabola.

Maximum and minimum points.

The maximum or minimum of the function is always obtained at the vertex, the following method is another derivation of the same fact using calculus, the advantage of this method is that it works for more general functions.

Taking $f(x) = ax^2 + bx + c$ as sample quadratic equation, to find its maximum or minimum points (which depends on a, if a > 0, it has a minimum point, if a < 0, it has a maximum point) we have to first, take its derivative:

$$f(x) = ax^2 + bx + c \Leftrightarrow f'(x) = 2ax + b$$

Then, we find the roots of $f'(x)$:

$$2ax + b = 0 \Rightarrow 2ax = -b \Rightarrow x = -\frac{b}{2a}$$

So, $-\frac{b}{2a}$ is the value x of f(x). Now, to find the y value, we substitute $x = -\frac{b}{2a}$ on f(x):

$$y = a\left(-\frac{b}{2a}\right)^2 + b\left(-\frac{b}{2a}\right) + c \Rightarrow y = \frac{ab^2}{4a^2} - \frac{b^2}{2a} + c \Rightarrow y = \frac{b^2}{4a} - \frac{b^2}{2a} + c \Rightarrow$$

$$y = \frac{b^2 - 2b^2 + 4ac}{4a} \Rightarrow y = \frac{-b^2 + 4ac}{4a} \Rightarrow y = -\frac{(b^2 - 4ac)}{4a} \Rightarrow y = -\frac{\Delta}{4a}$$

Thus, the maximum or minimum point coordinates are:

$$\left(-\frac{b}{2a}, -\frac{\Delta}{4a}\right).$$

Indeterminate Form

In calculus and other branches of mathematical analysis, an indeterminate form is an algebraic expression obtained in the context of limits.

Limits involving algebraic operations are often performed by replacing subexpressions by their limits; if the expression obtained after this substitution does not give enough information to determine the original limit, it is known as an indeterminate form. The indeterminate forms include 0^0, 0/0, 1^∞, ∞ ∞ ∞, ∞/∞, 0 × ∞, and ∞^0.

Discussion

The most common example of an indeterminate form is 0/0. As x approaches 0, the ratios x/x^3, x/x, and x^2/x go to ∞, 1, and 0 respectively. In each case, however, if the limits of the numerator and denominator are evaluated and plugged into the division operation, the resulting expression is 0/0.

So (roughly speaking) 0/0 can be 0 or it can be and, in fact, it is possible to construct similar examples converging to any particular value.

That is why the expression 0/0 is indeterminate.

More formally, the fact that the functions f and g both approach 0 as x approaches some limit point c is not enough information to evaluate the limit

$$\lim_{x \to c} \frac{f(x)}{g(x)}.$$

That limit could converge to any number, or diverge to infinity, or might not exist, depending on what the functions f and g are.

In some theories a value may be defined even where the function is discontinuous. For example $|x|/x$ is undefined for $x = 0$ in real analysis. However it is the sign function with sgn(0) = 0 when considering Fourier series or hyperfunctions.

Not every undefined algebraic expression is an indeterminate form. For example, the expression 1/0 is undefined as a real number but is not indeterminate. This is because any limit that gives rise to this form will diverge to infinity.

An expression representing an indeterminate form may sometimes be given a numerical value in settings other than the computation of limits. The expression 0^0 is defined as 1 when it represents an empty product. In the theory of power series, it is also often treated as 1 by convention, to make certain formulas more concise. (See the section "Zero to the zero power" in the article on exponentiation.) In the context of measure theory, it is usual to take $0.\infty$ to be 0.

Some Examples and None-xamples

The form 0/0

The indeterminate form 0/0 is particularly common in calculus because it often arises in the evaluation of derivatives using their limit definition.

As mentioned above,

$$\lim_{x \to 0} \frac{x}{x} = 1,$$

while

$$\lim_{x \to 0} \frac{x^2}{x} = 0.$$

This is enough to show that 0/0 is an indeterminate form. Other examples with this indeterminate form include

$$\lim_{x\to 0}\frac{\sin(x)}{x}=1$$

and

$$\lim_{x\to 49}\frac{x-49}{\sqrt{x}-7}=14.$$

Direct substitution of the number that x approaches into any of these expressions leads to the indeterminate form 0/0, but the limits take many different values. In fact, any desired value A can be obtained for this indeterminate form as follows:

$$\lim_{x\to 0}\frac{Ax}{x}=A.$$

Furthermore, the value infinity can also be obtained (in the sense of divergence to infinity):

$$\lim_{x\to 0}\frac{x}{x^3}=\infty.$$

The form 0^0

The indeterminate form 0^0 has been discussed since at least 1834. [1] The following examples illustrate that the form is indeterminate:

$$\lim_{x\to +0}x^0=1,$$

$$\lim_{x\to +0}0^x=0.$$

Thus, in general, knowing that $\lim_{x\to c} f(x) = 0^+$ and $\lim_{x\to c} g(x) = 0$ is not sufficient to calculate the limit

$$\lim_{x\to c}f(x)^{g(x)}.$$

If the functions f and g are analytic and f is not identically zero in a neighbourhood of c on the complex plane, then the limit of $f(z)^{g(z)}$ will always be 1. This also holds for real functions, but f must not be negative in the domain of the limit; alternatively, f can be the absolute value of an analytic function.

If the pair $(f(x), g(x))$ remains between two lines $y = mx$ and $y = Mx$, where m and M are positive numbers, as x approaches c and $f(x)$ and $g(x)$ approache 0, then the limit is always 1.

In many settings other than when evaluating limits 0^0 is taken to be defined as 1 even though it is an indeterminate form; see the section zero to the zero power in the article on exponentiation. One justification for this is provided by the preceding result. Another is that in power series, such as

$$e^x = \sum_{n=0}^{\infty} \frac{x^n}{n!},$$

when $x = 0$, then the term in which $n = 0$ has the correct value only if $0^0 = 1$. Yet another is that in combinatorial problems one must sometimes take 0^0 to be an empty product.

Undefined Forms That are not Indeterminate

The expression 1/0 is *not* an indeterminate form because there is no range of *distinct* values that f/g could approach. Specifically, if f approaches 1 and g approaches 0, then $|f/g|$ must diverge to infinity.

Notice that although f and g may be chosen (on an appropriate domain) so that f/g approaches either positive or negative infinity (in the sense of the extended real numbers), this variation does not create an indeterminate form (from one point of view, because they both diverge; from another point of view, because all infinities are equivalent in the real projective line).

Similarly, the expressions and are not indeterminate because any limit that gives rise to one of these forms will converge to 0 or diverge to infinity, respectively.

Evaluating Indeterminate Forms

The indeterminate nature of a limit's form does *not* imply that the limit does not exist, as many of the examples above show. In many cases, algebraic elimination, L'Hôpital's rule, or other methods can be used to manipulate the expression so that the limit can be evaluated.

For example, the expression x^2/x can be simplified to x at any point other than $x = 0$.

Thus, the limit of this expression as x approaches 0 (which depends only on points *near* 0, not at $x = 0$ itself) is the limit of x, which is 0. Most of the other examples above can also be evaluated using algebraic simplification.

L'Hôpital's rule is a general method for evaluating the indeterminate forms 0/0 and ∞/∞. This rule states that (under appropriate conditions)

$$\lim_{x\to c}\frac{f(x)}{g(x)}=\lim_{x\to c}\frac{f'(x)}{g'(x)},$$

where f' and g' are the derivatives of f and g. (Note that this rule does *not* apply to forms like "/0, 1/0, and so on; but these forms are not indeterminate either.) With luck, these derivatives will allow one to perform algebraic simplification and eventually evaluate the limit.

L'Hôpital's rule can also be applied to other indeterminate forms, using first an appropriate algebraic transformation. For example, to evaluate the form 0^0:

$$\ln\lim_{x\to c}f(x)^{g(x)}=\lim_{x\to c}\frac{\ln f(x)}{1/g(x)}.$$

The right-hand side is of the form "/", so L'Hôpital's rule applies to it. Notice that this equation is valid (as long as the right-hand side is defined) because the natural logarithm (ln) is a continuous function; it's irrelevant how well-behaved f and g may (or may not) be as long as f is asymptotically positive.

Although L'Hôpital's rule applies both to 0/0 and to "/", one of these may be better than the other in a particular case (because of the possibilities for algebraic simplification afterwards). You can change between these forms, if necessary, by transforming f/g to $(1/g)/(1/f)$.

4

Elementary Matrix

In mathematics, an elementary matrix is a simple matrix which differs from the identity matrix in a minimal way. The elementary matrices generate the general linear group of invertible matrices, and left (respectively, right) multiplication by an elementary matrix represent elementary row operations (respectively, elementary column operations).

In algebraic K-theory, "elementary matrices" refers *only* to the row-addition matrices.

Use in Solving Systems of Equations

Elementary *row* operations do not change the solution set of the system of linear equations represented by a matrix, and are used in Gaussian elimination (respectively, Gauss-Jordan elimination) to reduce a matrix to row echelon form (respectively, reduced row echelon form). The acronym "ero" is commonly used for "elementary row operations".

Elementary row operations do not change the kernel of a matrix (and hence do not change the solution set), but they *do* change the image.

Dually, elementary *column* operations do not change the *image*, but they do change the *kernel*. There are three types of Elementary matrices (they are nxn)

1) Permutation Matrix
2) Diagonal Matrix
3) Unipotent Matrix.

Operations

There are three types of elementary matrices, which correspond to three types of row operations (respectively, column operations):

Row Switching

A row within the matrix can be switched with another row.

$$R_i \leftrightarrow R_j$$

Row Multiplication

Each element in a row can be multiplied by a non-zero constant.

$$kR_i \rightarrow R_i, \text{ where } k \neq 0$$

Row Addition

A row can be replaced by the sum of that row and a multiple of another row.

$$R_i + kR_j \rightarrow R_i$$

Row-switching Transformations

This transformation, T_{ij}, switches all matrix elements on row i with their counterparts on row j.

The matrix resulting in this transformation is obtained by swapping row i and row j of the identity matrix.

$$T_{i,j} = \begin{bmatrix} 1 & & & & & & \\ & \ddots & & & & & \\ & & 0 & & 1 & & \\ & & & \ddots & & & \\ & & 1 & & 0 & & \\ & & & & & \ddots & \\ & & & & & & 1 \end{bmatrix}$$

That is, T_{ij} is the matrix produced by exchanging row i and row j of the identity matrix.

Properties

- The inverse of this matrix is itself: $T_{ij}^{-1}=T_{ij}$.
- Since the determinant of the identity matrix is unity, $\det[T_{ij}] = -1$.

  It follows that for any conformable square matrix A: $\det[T_{ij}A] = -\det[A]$.

Row-multiplying Transformations

$$T_i(m) = \begin{bmatrix} 1 & & & & & & \\ & \ddots & & & & & \\ & & 1 & & & & \\ & & & m & & & \\ & & & & 1 & & \\ & & & & & \ddots & \\ & & & & & & 1 \end{bmatrix}$$

This transformation, $T_i(m)$, multiplies all elements on row i by m where m is non zero. The matrix resulting in this transformation is obtained by multiplying all elements of row i of the identity matrix by m.

Properties

- The inverse of this matrix is: $T_i(m)^{-1} = T_i(1/m)$.
- The matrix and its inverse are diagonal matrices.
- $\det[T_i(\mathrm{m})] = m$. Therefore for a conformable square matrix A: $\det[T_i(m)A] = m \det[A]$.

Row-addition Transformations

This transformation, $T_{ij}(m)$, subtracts row j multiplied by m from row i. The matrix resulting in this transformation is obtained by taking row j of the identity matrix, and subtracting from it m times row i.

$$T_{i,j}(m) = \begin{bmatrix} 1 & & & & & & \\ & \ddots & & & & & \\ & & 1 & & & & \\ & & & \ddots & & & \\ & & m & & 1 & & \\ & & & & & \ddots & \\ & & & & & & 1 \end{bmatrix}$$

These are also called shear mappings or transvections.

Properties

- $T_{ij}(m)^{-1} = T_{ij}(-m)$ (inverse matrix).
- The matrix and its inverse are triangular matrices.

- $\det[T_{ij}(m)] = 1$. Therefore, for a conformable square matrix A: $\det[T_{ij}(1)A] = \det[A]$.

Row-addition transforms satisfy the Steinberg relations.

Eigenvalue, Eigenvector and Eigenspace

In mathematics, eigenvalue, eigenvector, and eigenspace are related concepts in the field of linear algebra. Linear algebra studies linear transformations, which are represented by matrices acting on vectors. Eigenvalues, eigenvectors and eigenspaces are properties of a matrix. They are computed by a method, give important information about the matrix, and can be used in matrix factorization. They have applications in areas of applied mathematics as diverse as finance and quantum mechanics.

In general, a matrix acts on a vector by changing both its magnitude and its direction. However, a matrix may act on certain vectors by changing only their magnitude, and leaving their direction unchanged (or, possibly, reversing it). These vectors are the eigenvectors of the matrix.

A matrix acts on an eigenvector by multiplying its magnitude by a factor, which is positive if its direction is unchanged and negative if its direction is reversed. This factor is the eigenvalue associated with that eigenvector. An eigenspace is the set of all eigenvectors that have the same eigenvalue. The concepts cannot be formally defined without prerequisites, including an understanding of matrices, vectors, and linear transformations.

Mathematical Definition

In linear algebra, there are two kinds of objects, "scalars", which are just numbers, and "vectors", which can be thought of as arrows, and which have both magnitude and direction. In place of the ordinary functions of algebra, the most important functions in linear algebra are called "linear transformations", and a linear transformation is usually given by a "matrix", an array of numbers. Thus instead of writing $f(x)$ we write M(v) where M is a matrix and v is a vector. The rules for using a matrix to transform a vector are given in the article linear algebra.

If the action of a matrix on a (nonzero) vector changes its magnitude but not its direction, then the vector is called an eigenvector of that matrix. A vector which is "flipped" to point in the opposite direction is also considered an eigenvector. Each eigenvector is, in effect, multiplied by a scalar, called the eigenvalue corresponding to that eigenvector. The eigenspace corresponding to one eigenvalue of a given matrix is the set of all eigenvectors of the matrix with that eigenvalue.

Many kinds of mathematical objects can be treated as vectors: ordered pairs, functions, harmonic modes, quantum states, and frequencies, are examples. In these cases, the concept of *direction* loses its ordinary meaning, and is given an abstract definition. Even so, if this abstract *direction* is unchanged by a given linear transformation, the prefix "eigen" is used, as in *eigenfunction, eigenmode, eigenstate,* and *eigenfrequency.*

Example

If a matrix is a diagonal matrix, then its eigenvalues are the numbers on the diagonal and its eigenvectors are basis vectors to which those numbers refer. For example, the matrix stretches every vector to three times its original length in the x-direction and shrinks every vector to half its original length in the y-direction.

Eigenvectors corresponding to the eigenvalue 3 are any multiple of the basis vector [1, 0]; together they constitute the eigenspace corresponding to the eigenvalue 3. Eigenvectors corresponding to the eigenvalue 0.5 are any multiple of the basis vector [0, 1]; together they constitute the eigenspace corresponding to the eigenvalue 0.5. In contrast, any other vector, [2, 8] for example, will change direction. The angle [2, 8] makes with the x-axis has tangent 4, but after being transformed, [2, 8] is changed to [6, 4], and the angle that vector makes with the x-axis has tangent 2/3.

$$\begin{bmatrix} 3 & 0 \\ 0 & 0.5 \end{bmatrix}$$

History

Eigenvalues are often introduced in the context of linear algebra or matrix theory. Historically, however, they arose in the study of quadratic forms and differential equations.

Euler studied the rotational motion of a rigid body and discovered the importance of the principal axes. Lagrange realized that the principal axes are the eigenvectors of the inertia matrix. In the early 19th century, Cauchy saw how their work could be used to classify the quadric surfaces, and generalized it to arbitrary dimensions. Cauchy also coined the term *racine caractéristique* (characteristic root) for what is now called *eigenvalue*; his term survives in *characteristic equation*.

Fourier used the work of Laplace and Lagrange to solve the heat equation by separation of variables in his famous 1822 book *Théorie analytique de la chaleur*. Sturm developed Fourier's ideas further and brought them to the attention of Cauchy, who combined them with his own ideas and arrived at the fact that symmetric matrices have real eigenvalues.

This was extended by Hermite in 1855 to what are now called Hermitian matrices.

Around the same time, Brioschi proved that the eigenvalues of orthogonal matrices lie on the unit circle, and Clebsch found the corresponding result for skew-symmetric matrices.

Finally, Weierstrass clarified an important aspect in the stability theory started by Laplace by realizing that defective matrices can cause instability.

In the meantime, Liouville studied eigenvalue problems similar to those of Sturm; the discipline that grew out of their work is now called *Sturm-Liouville theory*. Schwarz studied the first eigenvalue of Laplace's equation on general domains towards the end of the 19th century, while Poincaré studied Poisson's equation a few years later.

At the start of the 20th century, Hilbert studied the eigenvalues of integral operators by viewing the operators as infinite matrices. He was the first to use the German word *eigen* to denote eigenvalues and eigenvectors in 1904, though he may have been following a related usage by Helmholtz.

"Eigen" can be translated as "own", "peculiar to", "characteristic", or "individual"—emphasizing how important eigenvalues are to defining the unique nature of a specific transformation. For some time, the standard term in English was "proper value", but the more distinctive term "eigenvalue" is standard today.

The first numerical algorithm for computing eigenvalues and eigenvectors appeared in 1929, when Von Mises published the power method. One of the most popular methods today, the QR algorithm, was proposed independently by John G.F. Francis and Vera Kublanovskaya in 1961.

Technical Definition

Linear transformations of a vector space, such as rotation, reflection, stretching, compression, shear or any combination of these, may be visualized by the effect they produce on vectors. In other words, they are vector functions.

More formally, in a vector space L, a vector function A is defined if for each vector x of L there corresponds a unique vector y = A(x) of L. For the sake of brevity, the parentheses around the vector on which the transformation is acting are often omitted.

A vector function A is *linear* if it has the following two properties:

- *Additivity*: $A(x + y) = Ax + Ay$
- *Homogeneity*: $A(áx) = áAx$.

where x and y are any two vectors of the vector space L and á is any scalar. Such a function is variously called a *linear transformation*, *linear operator*, or *linear endomorphism* on the space L.

Given a linear transformation A, a non-zero vector x is defined to be an *eigenvector* of the transformation if it satisfies the eigenvalue equation $Ax = \lambda x$ for some scalar λ. In this situation, the scalar λ is called an *eigenvalue* of A corresponding to the eigenvector x.

The key equation in this definition is the eigenvalue equation, $Ax = \lambda x$. That is to say that the vector x has the property that its direction is not changed by the transformation A, but that it is only scaled by a factor of λ. Most vectors x will not satisfy such an equation: a typical vector x changes direction when acted on by A, so that Ax is not a multiple of x.

This means that only certain special vectors x are eigenvectors, and only certain special scalars λ are eigenvalues. Of course, if A is a multiple of the identity matrix, then no vector changes direction, and all non-zero vectors are eigenvectors.

The requirement that the eigenvector be non-zero is imposed because the equation $A0 = \lambda 0$ holds for every A and every λ. Since the equation is always trivially true, it is not an interesting case.

In contrast, an eigenvalue can be zero in a nontrivial way. Each eigenvector is associated with a specific eigenvalue. One eigenvalue can be associated with several or even with an infinite number of eigenvectors.

Geometrically, the eigenvalue equation means that under the transformation A eigenvectors experience only changes in magnitude and sign—the direction of Ax is the same as that of x. The eigenvalue λ is simply the amount of "stretch" or "shrink" to which a vector is subjected when transformed by A. If $\lambda = 1$, the vector remains unchanged (unaffected by the transformation).

A transformation I under which a vector x remains unchanged, Ix = x, is defined as identity transformation. If λ = d1, the vector flips to the opposite direction; this is defined as reflection.

If x is an eigenvector of the linear transformation A with eigenvalue λ, then any scalar multiple áx is also an eigenvector of A with the same eigenvalue. Similarly if more than one eigenvector shares the same eigenvalue λ, any linear combination of these eigenvectors will itself be an eigenvector with eigenvalue λ. Together with the zero vector, the eigenvectors of A with the same eigenvalue form a linear subspace of the vector space called an *eigenspace.*

The eigenvectors corresponding to different eigenvalues are linearly independent meaning, in particular, that in an n-dimensional space the linear transformation A cannot have more than n eigenvectors with different eigenvalues.

If a basis is defined in vector space, all vectors can be expressed in terms of components. For finite dimensional vector spaces with dimension n, linear transformations can be represented with $n \times n$ square matrices. Conversely, every such square matrix corresponds to a linear transformation for a given basis.

Thus, in a two-dimensional vector space R^2 fitted with standard basis, the eigenvector equation for a linear transformation A can be written in the following matrix representation: where the juxtaposition of matrices denotes matrix multiplication.

$$\begin{bmatrix} a_{11} & q_{12} \\ a_{21} & a_{22} \end{bmatrix} \begin{bmatrix} x \\ y \end{bmatrix} = \begin{bmatrix} x \\ y \end{bmatrix},$$

Left and Right Eigenvectors

The word eigenvector formally refers to the right eigenvector x_R. It is defined by the above eigenvalue equation $Ax_R = \lambda_R x_R$, and is the most commonly used eigenvector. However, the left eigenvector x_L exists as well, and is defined by $x_L A = \lambda_L x_L$.

Computation of Eigenvalues, and the Characteristic Equation

When a transformation is represented by a square matrix A, the eigenvalue equation can be expressed as:

$$A\mathrm{x} - \lambda I\mathrm{x} = 0.$$

This can be rearranged to

$$(A - \lambda I)\mathrm{x} = 0.$$

If there exists an inverse

$$(A - \lambda I)^{-1},$$

then both sides can be left multiplied by the inverse to obtain the trivial solution: x = 0. Thus we require there to be no inverse by assuming from linear algebra that the determinant equals zero:

$$\det(A \text{ d } \lambda I) = 0.$$

The determinant requirement is called the *characteristic equation* (less often, secular equation) of A, and the left-hand side is called the *characteristic polynomial.* When expanded, this gives a polynomial equation for λ. The eigenvector x or its components are not present in the characteristic equation.

Example

The matrix

$$\begin{bmatrix} 2 & 1 \\ 1 & 2 \end{bmatrix}$$

defines a linear transformation of the real plane. The eigenvalues of this transformation are given by the characteristic equation

$$\det\begin{bmatrix} 2-\lambda & 1 \\ 1 & 2-\lambda \end{bmatrix} = 0,$$

$$\Rightarrow (2-\lambda)^2 - 1 = 0,$$

$$\Rightarrow \lambda^2 - 4\lambda + 3 = 0.$$

The roots of this equation (i.e. the values of λ for which the equation holds) are λ = 1 and λ = 3. Having found the eigenvalues, it is possible to find the eigenvectors. Considering first the eigenvalue λ = 3, we have

$$\begin{bmatrix} 2 & 1 \\ 1 & 2 \end{bmatrix}\begin{bmatrix} x \\ y \end{bmatrix} = 3\begin{bmatrix} x \\ y \end{bmatrix}.$$

After matrix-multiplication $\begin{bmatrix} 2x+y \\ x+2y \end{bmatrix} = 3\begin{bmatrix} x \\ y \end{bmatrix}$.

This matrix equation represents a system of two linear equations $2x + y = 3x$ and $x + 2y = 3y$.

Both the equations reduce to the single linear equation $x = y$. To find an eigenvector, we are free to choose any value for x (except 0), so by picking x=1 and setting y=x, we find an eigenvector with eigenvalue 3 to be

$$\begin{bmatrix} 1 \\ 1 \end{bmatrix}.$$

We can confirm this is an eigenvector with eigenvalue 3 by checking the action of the matrix on this vector:

$$\begin{bmatrix} 2 & 1 \\ 1 & 2 \end{bmatrix}\begin{bmatrix} 1 \\ 1 \end{bmatrix} = \begin{bmatrix} 3 \\ 3 \end{bmatrix} = 3\begin{bmatrix} 1 \\ 1 \end{bmatrix}.$$

Any scalar multiple of this eigenvector will also be an eigenvector with eigenvalue 3.

For the eigenvalue $\lambda = 1$, a similar process leads to the equation $x = \text{d}\ y$, and hence an eigenvector with eigenvalue 1 is given by

$$\begin{bmatrix} 1 \\ -1 \end{bmatrix}.$$

The complexity of the problem for finding roots/eigenvalues of the characteristic polynomial increases rapidly with increasing the degree of the polynomial (the dimension of the vector space). There are exact solutions for dimensions below 5, but for dimensions greater than or equal to 5 there are generally no exact solutions and one has to resort to numerical methods to find them approximately.

Worse, this computational procedure can be very inaccurate in the presence of round-off error, because the roots of a polynomial are an extremely sensitive function of the coefficients.

Efficient, accurate methods to compute eigenvalues and eigenvectors of arbitrary matrices were not known until the advent of the QR algorithm in 1961.

For large Hermitian sparse matrices, the Lanczos algorithm is one example of an efficient iterative method to compute eigenvalues and eigenvectors, among several other possibilities.

Existence and Multiplicity of Eigenvalues

For transformations on real vector spaces, the coefficients of the characteristic polynomial are all real. However, the roots are not necessarily real; they may include complex numbers with a non-zero imaginary component.

For example, a matrix representing a planar rotation of 45 degrees will not leave any non-zero vector pointing in the same direction. Over a complex vector space, the fundamental theorem of algebra guarantees that the characteristic polynomial has at least one root, and thus the linear transformation has at least one eigenvalue.

As well as distinct roots, the characteristic equation may also have repeated roots. However, having repeated roots does not imply there are multiple distinct (i.e., linearly independent) eigenvectors with that eigenvalue. The *algebraic multiplicity* of an eigenvalue is defined as the multiplicity of the corresponding root of the characteristic polynomial.

The geometric multiplicity of an eigenvalue is defined as the dimension of the associated eigenspace, i.e. number of linearly independent eigenvectors with that eigenvalue.

Over a complex space, the sum of the algebraic multiplicities will equal the dimension of the vector space, but the sum of the geometric multiplicities may be smaller.

In a sense, then it is possible that there may not be sufficient eigenvectors to span the entire space. This is intimately related to the question of whether a given matrix may be diagonalized by a suitable choice of coordinates.

Shear

Horizontal shear. The shear angle ϕ is given by $k = \cot \phi$.

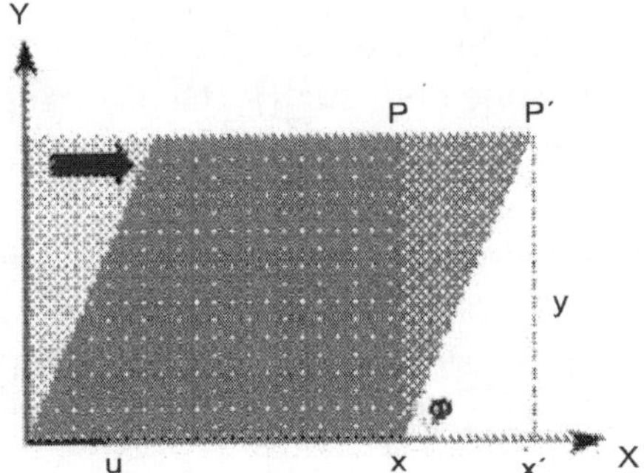

Shear in the plane is a transformation in which all points along a given line remain fixed while other points are shifted parallel to that line by a distance proportional to their perpendicular distance from the line. Shearing a plane figure does not change its area. Shear can be horizontal " along the X axis, or vertical " along the Y axis. In horizontal shear, a point P of the plane moves parallel to the X axis to the place P' so that its coordinate y does not change while the x coordinate increments to become $x' = x + k\,y$, where k is called the shear factor.

The matrix of a horizontal shear transformation is $\begin{bmatrix} 1 & k \\ 0 & 1 \end{bmatrix}$. The characteristic equation is $\lambda^2 - 2\,\lambda + 1 = (1 - \lambda)^2 = 0$ which has a single, repeated root $\lambda = 1$. Therefore, the eigenvalue $\lambda = 1$ has algebraic multiplicity 2. The eigenvector(s) are found as solutions of

$$\begin{bmatrix} 1-1 & -k \\ 0 & 1-1 \end{bmatrix}\begin{bmatrix} x \\ y \end{bmatrix} = \begin{bmatrix} 0 & -k \\ 0 & 0 \end{bmatrix}\begin{bmatrix} x \\ y \end{bmatrix} = -ky = 0$$

The last equation is equivalent to $y = 0$, which is a straight line along the x axis. This line represents the one-dimensional eigenspace. In the case of shear the algebraic multiplicity of the eigenvalue (2) is greater than its geometric multiplicity (1, the dimension of the eigenspace). The eigenvector is a vector along the x axis. The case of vertical shear with transformation matrix $\begin{bmatrix} 1 & 0 \\ k & 1 \end{bmatrix}$ is dealt with in a similar way; the eigenvector in vertical shear is along the y axis. Applying repeatedly the shear transformation changes the direction of any vector in the plane closer and closer to the direction of the eigenvector.

Uniform Scaling and Reflection

As a one-dimensional vector space, consider a rubber string tied to unmoving support in one end, such as that on a child's sling. Pulling the string away from the point of attachment stretches it and elongates it by some scaling factor λ which is a real number. Each vector on the string is stretched equally, with the same scaling factor λ, and although elongated, it preserves its original direction. For a two-dimensional vector space, consider a rubber sheet stretched equally in all directions such as a small area of the surface of an inflating balloon.

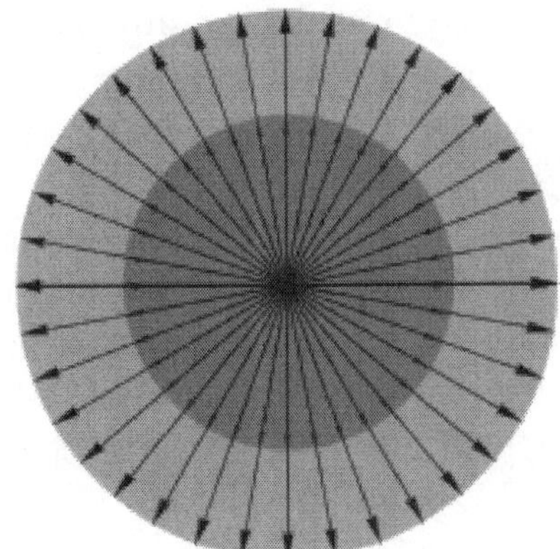

Figure: *When a surface is stretching equally in all directions (a homothety) any one of the radial vectors can be the eigenvector.*

All vectors originating at the fixed point on the balloon surface (the origin) are stretched equally with the same scaling factor λ. This transformation in two-dimensions is described by the 2×2 square matrix:

$$Ax = \begin{bmatrix} \lambda & 0 \\ 0 & \lambda \end{bmatrix}\begin{bmatrix} x \\ y \end{bmatrix} = \begin{bmatrix} \lambda.x + 0.y \\ 0.x + \lambda.y \end{bmatrix} = \lambda\begin{bmatrix} x \\ y \end{bmatrix} = \lambda x.$$

Expressed in words, the transformation is equivalent to multiplying the length of *any* vector by λ while preserving its original direction. Since the vector taken was arbitrary, every non-zero vector in the vector space is an eigenvector. Whether the transformation is stretching (elongation, extension, inflation), or shrinking (compression, deflation) depends on the scaling factor: if $\lambda > 1$, it is stretching; if $\lambda < 1$, it is shrinking. Negative values of λ correspond to a reversal of direction, followed by a stretch or a shrink, depending on the absolute value of λ.

Unequal Scaling

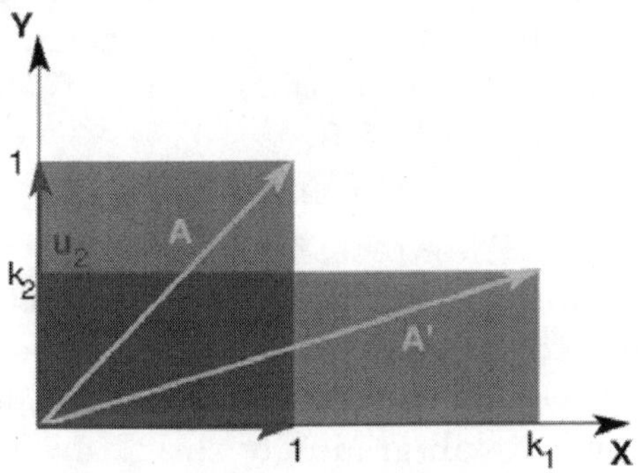

Vertical shrink ($k_2 < 1$) and horizontal stretch ($k_1 > 1$) of a unit square. Eigenvectors are u_1 and u_2 and eigenvalues are $\lambda_1 = k_1$ and $\lambda_2 = k_2$. This transformation orients all vectors towards the principal eigenvector u_1.

For a slightly more complicated example, consider a sheet that is stretched unequally in two perpendicular directions along the coordinate axes, or, similarly, stretched in one direction, and shrunk in the other direction. In this case, there are two different scaling factors: k_1 for the scaling in direction x, and k_2 for the scaling in direction y. The transformation matrix is

$$\begin{bmatrix} k_1 & 0 \\ 0 & k_2 \end{bmatrix},$$

and the characteristic equation is $(k_1 - \lambda)(k_2 - \lambda) = 0$. The eigenvalues, obtained as roots of this equation are $\lambda_1 = k_1$, and $\lambda_2 = k_2$ which means, as expected, that the two eigenvalues are the scaling factors in the two directions. Plugging k_1 back in the eigenvalue equation gives one of the eigenvectors:

$$\begin{bmatrix} 0 & 0 \\ 0 & k_2 - k_1 \end{bmatrix}\begin{bmatrix} x \\ y \end{bmatrix} = \begin{bmatrix} 0 \\ 0 \end{bmatrix} \text{ or, more simply, } y = 0.$$

Thus, the eigenspace is the x-axis. Similarly, substituting $\lambda = k_2$ shows that the corresponding eigenspace is the y-axis. In this case, both eigenvalues have algebraic and geometric multiplicities equal to 1. If a given eigenvalue is greater than 1, the vectors are stretched in the direction of the corresponding eigenvector; if less than 1, they are shrunken in that direction.

Negative eigenvalues correspond to reflections followed by a stretch or shrink. In general, matrices that are diagonalizable over the real numbers represent scalings and reflections: the eigenvalues represent the scaling factors (and appear as the diagonal terms), and the eigenvectors are the directions of the scalings. The figure shows the case where $k_1 > 1$ and $1 > k_2 > 0$.

The rubber sheet is stretched along the x axis and simultaneously shrunk along the y axis. After repeatedly applying this transformation of stretching/shrinking many times, almost any vector on the surface of the rubber sheet will be oriented closer and closer to the direction of the x axis (the direction of stretching). The exceptions are vectors along the y-axis, which will gradually shrink away to nothing.

Rotation

A rotation in a plane is a transformation that describes motion of a vector, plane, coordinates, etc., around a fixed point. Clearly, for rotations other than through 0° and 360°, every vector in the real plane will have its direction changed, and thus there cannot be any eigenvectors. But this is not necessarily true if we consider the same matrix over a complex vector space. A counterclockwise rotation in the horizontal plane about the origin at an angle ϕ is represented by the matrix

$$R = \begin{bmatrix} \cos\varphi & -\sin\varphi \\ \sin\varphi & \cos\varphi \end{bmatrix}.$$

The characteristic equation of R is $\lambda^2 - 2\lambda \cos \phi + 1 = 0$. This quadratic equation has a discriminant $D = 4 (\cos^2 \phi - 1) = - 4 \sin^2 \phi$ which is a negative number whenever ϕ is not equal a multiple of 180°.

A rotation of 0°, 360°, ... is just the identity transformation, (a uniform scaling by +1) while a rotation of 180°, 540°, ..., is a reflection (uniform scaling by -1). Otherwise, as expected, there are no real eigenvalues or eigenvectors for rotation in the plane.

Rotation Matrices on Complex Vector Spaces

The characteristic equation has two complex roots λ_1 and λ_2. If we choose to think of the rotation matrix as a linear operator on the complex two dimensional, we can consider these complex eigenvalues.

The roots are complex conjugates of each other: $\lambda_{1,2} = \cos \phi \pm i \sin \phi = e^{\pm i\phi}$, each with an algebraic multiplicity equal to 1, where i is the imaginary unit.

The first eigenvector is found by substituting the first eigenvalue, λ_1, back in the eigenvalue equation:

$$\begin{bmatrix} \cos\varphi - \lambda_1 & -\sin\varphi \\ \sin\varphi & \cos\varphi - \lambda_1 \end{bmatrix} \begin{bmatrix} x \\ y \end{bmatrix} = \begin{bmatrix} -i\sin\varphi & -\sin\varphi \\ \sin\varphi & -\sin\varphi \end{bmatrix} \begin{bmatrix} x \\ y \end{bmatrix} = \begin{bmatrix} 0 \\ 0 \end{bmatrix}.$$

The last equation is equivalent to the single equation $x = iy$, and again we are free to set $x = 1$ to give the eigenvector

$$\begin{bmatrix} 1 \\ -i \end{bmatrix}.$$

Similarly, substituting in the second eigenvalue gives the single equation $x = -iy$ and so the eigenvector is given by

$$\begin{bmatrix} 1 \\ i \end{bmatrix}.$$

Although not diagonalizable over the reals, the rotation matrix is diagonalizable over the complex numbers, and again the eigenvalues appear on the diagonal. Thus rotation matrices acting on complex spaces can be thought of as scaling matrices, with complex scaling factors.

Infinite-dimensional Spaces and Spectral Theory

If the vector space is an infinite dimensional Banach space, the notion of eigenvalues can be generalized to the concept of spectrum. The spectrum is the set of scalars λ for which $(T - \lambda I)^{-1}$ is not defined; that is, such that $T - \lambda I$ has no bounded inverse.

Clearly if λ is an eigenvalue of T, λ is in the spectrum of T. In general, the converse is not true. There are operators on Hilbert or Banach spaces which have no eigenvectors at all. This can be seen in the following example. The bilateral shift on the Hilbert space – $l^2(Z)$ (that is, the space of all sequences of scalars ... a_{-1}, a_0, a_1, a_2, ... such that

$$...+|a_{-1}|^2 + |a_0|^2 ++ |a_1|^2 ++ |a_2|^2 +...$$

converges) has no eigenvalue but does have spectral values.

In infinite-dimensional spaces, the spectrum of a bounded operator is always nonempty. This is also true for an unbounded self adjoint operator. Via its spectral measures, the spectrum of any self adjoint operator, bounded or otherwise, can be decomposed into absolutely continuous, pure point, and singular parts.

The hydrogen atom is an example where both types of spectra appear. The eigenfunctions of the hydrogen atom Hamiltonian are called eigenstates and are grouped into two categories. The bound states of the hydrogen atom correspond to the discrete part of the spectrum (they have a discrete set of eigenvalues which can be computed by Rydberg formula) while the ionization processes are described by the continuous part (the energy of the collision/ionization is not quantized).

Eigenfunctions

A common example of such maps on infinite dimensional spaces are the action of differential operators on function spaces. As an example, on the space of infinitely differentiable functions, the process of differentiation defines a linear operator since

$$\frac{d}{dt}(af+bg)=a\frac{df}{dt}+b\frac{dg}{dt},$$

where $f(t)$ and $g(t)$ are differentiable functions, and a and b are constants.

The eigenvalue equation for linear differential operators is then a set of one or more differential equations. The eigenvectors are commonly called eigenfunctions.

The most simple case is the eigenvalue equation for differentiation of a real valued function by a single real variable. In this case, the eigenvalue equation becomes the linear differential equation

$$\frac{d}{dx}f(x)=\lambda f(x).$$

Here λ is the eigenvalue associated with the function, $f(x)$. This eigenvalue equation has a solution for all values of λ. If λ is zero, the solution is

$$f(x)=A,$$

where A is any constant; if λ is non-zero, the solution is the exponential function

$$f(x)=Ae^{\lambda x},$$

If we expand our horizons to complex valued functions, the value of λ can be any complex number. The spectrum of d/dt is therefore the whole complex plane. This is an example of a continuous spectrum.

Waves on a String

The shape of a standing wave in a string fixed at its boundaries is an example of an eigenfunction of a differential operator. The admittable eigenvalues are governed by the length of the string and determine the frequency of oscillation

The displacement, $h(x,t)$, of a stressed rope fixed at both ends, like the vibrating strings of a string instrument, satisfies the wave equation

$$\frac{\partial^2 h}{\partial t^2} = c^2 \frac{\partial^2 h}{\partial x^2};$$

which is a linear partial differential equation, where c is the constant wave speed. The normal method of solving such an equation is separation of variables. If we assume that h can be written as the product of the form $X(x)T(t)$, we can form a pair of ordinary differential equations:

$$X'' = -\frac{\omega^2}{c^2}X \text{ and } T'' = -\omega^2 T.$$

Each of these is an eigenvalue equation (the unfamiliar form of the eigenvalue is chosen merely for convenience). For any values of the eigenvalues, the eigenfunctions are given by

$$X = \sin\left(\frac{\omega x}{c} + \phi\right) \textit{ and } T = \sin(\omega t + \psi).$$

If we impose boundary conditions (that the ends of the string are fixed with $X(x)=0$ at $x=0$ and $x=L$, for example) we can constrain the eigenvalues. For those boundary conditions, we find

$$\sin(\phi) = 0,$$

and so the phase angle $\phi = 0$ and

$$\sin\left(\frac{\omega L}{c}\right) = 0.$$

Thus, the constant ω is constrained to take one of the values

$$\omega_n = \frac{nc\pi}{L},$$

where n is any integer. Thus the clamped string supports a family of standing waves of the form

$$h(x,t) = \sin(n\pi x / L)\sin(\omega_n t).$$

From the point of view of our musical instrument, the frequency ω_n is the frequency of the nth harmonic overtone.

Eigende Composition

The spectral theorem for matrices can be stated as follows. Let A be a square $n \times n$ matrix. Let $q_1 \ldots q_k$ be an eigenvector basis, i.e. an indexed set of k linearly independent eigenvectors, where k is the dimension of the space spanned by the eigenvectors of A.

If $k = n$, then A can be written

$$A = Q\Lambda Q^{-1}$$

where Q is the square $n \times n$ matrix whose i-th column is the basis eigenvector q_i of A and K is the diagonal matrix whose diagonal elements are the corresponding eigenvalues, i.e. $\ddot{E}_{ii} = \lambda_i$.

Applications

Schrodinger Equation

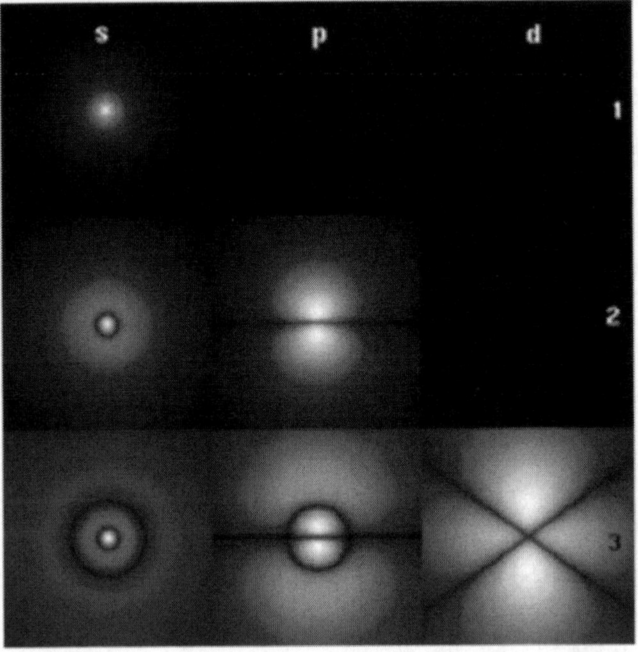

The wavefunctions associated with the bound states of an electron in a hydrogen atom can be seen as the eigenvectors of the hydrogen atom Hamiltonian as well as of the angular momentum operator. They are associated with eigenvalues interpreted as their energies (increasing downward: n=1,2,3,...) and angular momentum (increasing across: s, p, d,...). The illustration shows the square of the absolute value of the wavefunctions. Brighter areas correspond to higher probability density for a position measurement. The centre of each figure is the atomic nucleus, a proton.

An example of an eigenvalue equation where the transformation T is represented in terms of a differential operator is the time-independent Schrödinger equation in quantum mechanics:

$$H\psi_E = E\psi_E$$

where H, the Hamiltonian, is a second-order differential operator and ψ_E, the wavefunction, is one of its eigenfunctions corresponding to the eigenvalue E, interpreted as its energy.

However, in the case where one is interested only in the bound state solutions of the Schrödinger equation, one looks for ψ_E within the space of square integrable functions.

Since this space is a Hilbert space with a well-defined scalar product, one can introduce a basis set in which ψ_E and H can be represented as a one-dimensional array and a matrix respectively. This allows one to represent the Schrödinger equation in a matrix form.

Bra-ket notation is often used in this context. A vector, which represents a state of the system, in the Hilbert space of square integrable functions is represented by $|\Psi_E\rangle$. In this notation, the Schrödinger equation is:

$$H|\Psi_E\rangle = E|\Psi_E\rangle$$

where $|\Psi_E\rangle$ is an eigenstate of H. It is a self adjoint operator, the infinite dimensional analog of Hermitian matrices. As in the matrix case, in the equation above $H|\Psi_E\rangle$ is understood to be the vector obtained by application of the transformation H to $|\Psi_E\rangle$.

Molecular Orbitals

In quantum mechanics, and in particular in atomic and molecular physics, within the Hartree-Fock theory, the atomic and molecular orbitals can be defined by the eigenvectors of the Fock operator. The corresponding eigenvalues are interpreted as ionization potentials via Koopmans' theorem.

In this case, the term eigenvector is used in a somewhat more general meaning, since the Fock operator is explicitly dependent on the orbitals and their eigenvalues.

If one wants to underline this aspect one speaks of nonlinear eigenvalue problem. Such equations are usually solved by an iteration procedure, called in this case self-consistent field method.

In quantum chemistry, one often represents the Hartree-Fock equation in a non-orthogonal basis set. This particular representation is a generalized eigenvalue problem called Roothaan equations.

Geology and Glaciology

In geology, especially in the study of glacial till, eigenvectors and eigenvalues are used as a method by which a mass of information of a clast fabric's constituents' orientation and dip can be summarized in a 3-D space by six numbers.

In the field, a geologist may collect such data for hundreds or thousands of clasts in a soil sample, which can only be compared graphically such as in a Tri-Plot (Sneed and Folk), or as a Stereonet on a Wulff Net. The output for the orientation tensor is in the three orthogonal (perpendicular) axes of space.

Eigenvectors output from programs such as Stereo32 are in the order $E_1 \geq E_2 \geq E_3$, with E_1 being the primary orientation of clast orientation/dip, E_2 being the secondary and E_3 being the tertiary, in terms of strength. The clast orientation is defined as the eigenvector, on a compass rose of 360°.

Dip is measured as the eigenvalue, the modulus of the tensor: this is valued from 0° (no dip) to 90° (vertical). The relative values of E_1, E_2, and E_3 are dictated by the nature of the sediment's fabric. If $E_1 = E_2 = E_3$, the fabric is said to be isotropic. If $E_1 = E_2 > E_3$ the fabric is planar. If $E_1 > E_2 > E_3$ the fabric is linear.

Factor Analysis

PCA of a multivariate Gaussian distribution centred at (1,3) with a standard deviation of 3 in roughly the (0.878, 0.478) direction and of 1 in the orthogonal direction.

In factor analysis, the eigenvectors of a covariance matrix or correlation matrix correspond to factors, and eigenvalues to the variance explained by these factors. Factor analysis is a statistical technique used in the social sciences and in marketing, product management, operations research, and other applied sciences that deal with large quantities of data.

The objective is to explain most of the covariability among a number of observable random variables in terms of a smaller number of unobservable latent variables called factors.

The observable random variables are modelled as linear combinations of the factors, plus unique variance terms. Eigenvalues are used in analysis used by Q-methodology software; factors with eigenvalues greater than 1.00 are considered significant, explaining an important amount of the variability in the data, while eigenvalues less than 1.00 are considered too weak, not explaining a significant portion of the data variability.

Vibration Analysis

Eigenvalue problems occur naturally in the vibration analysis of mechanical structures with many degrees of freedom. The eigenvalues are used to determine the natural frequencies of vibration, and the eigenvectors determine the shapes of these vibrational modes. The orthogonality properties of the eigenvectors allows decoupling of the differential equations so that the system can be represented as linear summation of the eigenvectors. The eigenvalue problem of complex structures is often solved using finite element analysis.

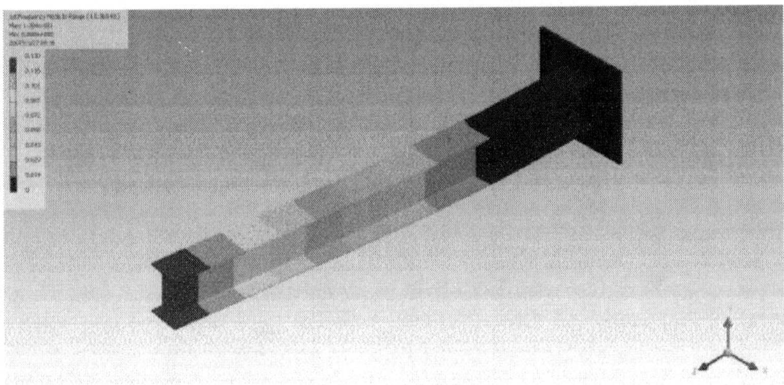

Eigenfaces

In image processing, processed images of faces can be seen as vectors whose components are the brightnesses of each pixel. The dimension of this vector space is the number of pixels. The eigenvectors of the covariance matrix associated with a large set of normalized pictures of faces are called eigenfaces; this is an example of principal components analysis. They are very useful for expressing any face image as a linear combination of some of them. In the facial recognition branch of biometrics, eigenfaces provide a means of applying data compression to faces for identification purposes. Research related to eigen vision systems determining hand gestures has also been made.

Similar to this concept, eigenvoices represent the general direction of variability in human pronunciations of a particular utterance, such as a word in a language. Based on a linear combination of such eigenvoices, a new voice pronunciation of the word can be constructed.

These concepts have been found useful in automatic speech recognition systems, for speaker adaptation.

Tensor of Inertia

In mechanics, the eigenvectors of the inertia tensor define the principal axes of a rigid body. The tensor of inertia is a key quantity required in order to determine the rotation of a rigid body around its centre of mass.

Stress Tensor

In solid mechanics, the stress tensor is symmetric and so can be decomposed into a diagonal tensor with the eigenvalues on the diagonal and eigenvectors as a basis. Because it is diagonal, in this orientation, the stress tensor has no shear components; the components it does have are the principal components.

Eigenvalues of a Graph

In spectral graph theory, an eigenvalue of a graph is defined as an eigenvalue of the graph's adjacency matrix A, or (increasingly) of the graph's Laplacian matrix, which is either $T'A$ or $I'T^{1/2}AT^{-1/2}$, where T is a diagonal matrix holding the degree of each vertex, and in $T^{-1/2}$, 0 is substituted for $0^{-1/2}$.

The kth principal eigenvector of a graph is defined as either the eigenvector corresponding to the kth largest eigenvalue of A, or the eigenvector corresponding to the kth smallest eigenvalue of the Laplacian. The first principal eigenvector of the graph is also referred to merely as the principal eigenvector.

The principal eigenvector is used to measure the centrality of its vertices. An example is Google's PageRank algorithm. The principal eigenvector of a modified adjacency matrix of the World Wide Web graph gives the page ranks as its components. This vector corresponds to the stationary distribution of the Markov chain represented by the row-normalized adjacency matrix; however, the adjacency matrix must first be modified to ensure a stationary distribution exists. The second smallest eigenvector can be used to partition the graph into clusters, via spectral clustering. Other methods are also available for clustering.

Introduction to Matrices

If you were asked for your weight in pounds, you would use a real number such as 140 to answer the question. If you were asked for your height in inches, you would answer with another real number such as 66.5. If we asked these questions to everyone in the class, we would want some way to know which weight goes with which height. One way to organize this data is to use an ordered pair.

We could represent your weight and height with the ordered pair (140, 66.5). This is called an ordered pair because we always list the information in the same order. In other words, we list weight first and then height in every pair of numbers, so (140, 66.5) would be different from (66.5, 140).

The elements are the individual pieces of information. Elements are also referred to as entries or components. In this book, we will only use real numbers as elements. The elements of this ordered pair are 140 and 66.5. We could also ask you for your age in years and append that information so that we have the ordered triple (140, 66.5, 18). We could ask you for n pieces of information, where n is any counting number.

If we arrange the n pieces of information in a specific order, we call it an ordered n-tuple. In general, lists of ordered information are called vectors. If we write them in rows, as we did above, we call them row vectors. If we write them in columns, such as

$$\begin{bmatrix} 140 \\ 66.5 \end{bmatrix} \text{ and } \begin{bmatrix} 140 \\ 66.5 \\ 18 \end{bmatrix},$$

we call them column vectors.

Definition 1: A real n-vector is an ordered n-tuple of real numbers.

The real numbers are called the elements of the vector.

Since we are only working with real numbers in this book, we will drop the word real when referring to vectors. When it is not important to specify how many elements are in the vector, we drop the qualifier n.

Remark 1: Did you notice that we used parentheses on some vectors and brackets on others? Actually, both are accepted notations, but we will use brackets for consistency throughout the rest of the book.

Remark 2: Sometimes you will see the elements of a row vector separated by commas. Commas are not necessary unless confusion can arise without the use of commas. If you were asked to add, subtract, or multiply real numbers, you would know what to do.

If we are going to use vectors to help us organize our information, we also need rules for vectors so that when we add, subtract, or multiply vectors, we get the same solutions as if we had not organized our data this way.

Remember that vectors are simply tools that we use to display information in an organized manner. Therefore, we do not want our solutions to change just because we organized our data into a vector. As we study this book, we will learn more about how to perform mathematical operations with vectors.

Consider the following information: The Cardinals win seven, lose six, and tie one. The Eagles win five, lose eight, and tie one. The Falcons win two, lose twelve, and have no ties. The Owls win nine, lose five, and have no ties.

We can represent this data using the four vectors $[7 \quad 6 \quad 1]$, $[5 \quad 8 \quad 1]$, $[2 \quad 12 \quad 0]$, and $[9 \quad 5 \quad 0]$. However, it would be nice if we could combine all these vectors together into one set of data. If we consider each vector as one row of an array, then we will have all our data in one arrangement.

Definition 2: A real matrix is an arrangement of real numbers into rows and columns.

The real numbers are called the elements of the matrix.

Since we are only working with real numbers in this book, we also will drop the word real when referring to matrices. Notice that a vector is a special matrix that has only one row or one column. When we organize our vectors into a matrix, it could look like this:

$$\begin{bmatrix} 7 & 6 & 1 \\ 5 & 8 & 1 \\ 2 & 12 & 0 \\ 9 & 5 & 0 \end{bmatrix}$$

Basically, we have all of our information organized into one arrangement called a matrix.

We have all the appropriate numbers in our matrix, but if we want to know which numbers correspond to which team, we have to look back at our paragraph. For this reason, we often label our matrices. Labels are not a formal part of the matrix, but they are very useful. Our matrix could look like this after it has been labelled:

$$\begin{array}{c} \\ C \\ E \\ F \\ O \end{array}\begin{array}{c} \begin{array}{ccc} W & L & T \end{array} \\ \begin{bmatrix} 7 & 6 & 1 \\ 5 & 8 & 1 \\ 2 & 12 & 0 \\ 9 & 5 & 0 \end{bmatrix} \end{array}$$

Just by looking at this matrix, we can tell that the Owls won the most games and the Falcons lost the most. One of the advantages of matrices is that information is easier to see and compare than when it is not organized into a matrix.

This matrix is referred to as a 4 by 3 matrix (often written 4x3) because there are 4 rows and 3 columns. Therefore, the dimensions of this matrix are 4 by 3. The dimensions of a matrix tell you the "size" of the matrix because they tell you the number of rows and columns in the matrix. By convention, we list the number of rows before the number of columns.

Definition 3: The dimensions of a matrix are the number of rows and columns (listed in that order) of the matrix.

Each element of the matrix is named according to its position. Typically, capital letters represent matrices and small letters with subscripts represent elements in the matrix. Since vectors can be considered to be matrices with only one row or one column, they could be labelled with capital letters also.

However, vectors are usually represented by small letters. If we name the above matrix A, the element 6 is in the position a_{12} (read a one two) because it is in row 1 and column 2.

Also by convention, we list the row number of the element before the column number. An element in row i and column j would be denoted by a_{ij}. This gives us a compact way to refer to specific elements of a matrix.

Remark 3: Although some mathematicians make a distinction between a 1 by 1 matrix, a 1-vector, and a real number, we will not make any distinction between them and will treat them exactly the same.

Can you represent the same information as before in a 3 by 4 matrix? Yes, you can. It would look like the matrix B which follows.

$$A = \begin{array}{c} \\ C \\ E \\ F \\ O \end{array} \begin{array}{c} \begin{array}{ccc} W & L & T \end{array} \\ \begin{bmatrix} 7 & 6 & 1 \\ 5 & 8 & 1 \\ 2 & 12 & 0 \\ 9 & 5 & 0 \end{bmatrix} \end{array} \qquad B = \begin{array}{c} \\ W \\ L \\ T \end{array} \begin{array}{c} \begin{array}{cccc} C & E & F & O \end{array} \\ \begin{bmatrix} 7 & 5 & 2 & 9 \\ 6 & 8 & 12 & 5 \\ 1 & 1 & 0 & 0 \end{bmatrix} \end{array}$$

Matrix B is the transpose of A, and A is the transpose of B. Transposing a matrix results in writing the columns as rows and the rows as columns, but what really happens is that element a_{ij} is placed in the position b_{ji} of the new matrix. Therefore, a_{12} moves to the position b_{21} when we form the transpose of A. The transpose of A is denoted by A^T (read A transpose). Therefore, matrix B is A^T.

Definition 4: By the transpose of the m by n matrix A, denoted by A^T, we mean the n by m matrix which has a_{ji} as its $(i,j)^{th}$ element.

Definition 5: We say that two m by n matrices, A and B are equal if their corresponding elements are equal.

In other words, $A = B$ if A and B have the same dimensions and $a_{11} = b_{11}$, $a_{12} = b_{12}$, etc. Is $A = A^T$? Usually not, but we have a special word for a matrix which satisfies $A = A^T$.

Definition 6: A matrix is said to be symmetric if $A = A^T$.

Observe that the following matrix is symmetric:

$$S = \begin{bmatrix} 9 & 2 & 5 & 1 \\ 2 & 7 & 0 & 8 \\ 5 & 0 & 4 & 6 \\ 1 & 8 & 6 & 3 \end{bmatrix}$$

Notice that $a_{ij} = a_{ji}$ for all i and j; as is true for all symmetric matrices. Symmetric matrices are easy to spot because if you draw a line down the main diagonal (from 9 to 3 in this matrix), then the two halves are mirror images of each other. Symmetric matrices have many special qualities that will be used when you study matrices in more detail.

The matrix S, given above, has another special property; it is a square matrix because S has the same number of rows as columns. Notice that S is a 4 by 4 square matrix. We said that the main diagonal for S runs from 9 to 3. For any square matrix, the main diagonal runs from the upper left corner to the lower right corner.

Definition 7: We say that an m by n matrix is square if $m = n$.

Addition of Matrices

If the Cardinals won 7 games in the first half of the regular season and won 8 in the second half, how many games did they win during the regular season? You know that the answer is 15 because 7 + 8 = 15.

The Eagles lost 8 games in the first half and lost 6 in the second half of the season. How many games did the Eagles lose all season? They lost 14 games. We know how to answer these questions using real numbers because we have represented our data by real numbers, and addition, subtraction, and multiplication are all defined and well-known operations for real numbers.

However, how would we add when our information is represented by matrices? Let the matrix A represent the statistics from the first half of the season, and let the matrix B represent the statistics from the second half of the season.

$$\begin{array}{c} \\ C \\ E \\ F \\ O \end{array}\begin{array}{c} \begin{array}{ccc} W & L & T \end{array} \\ \begin{bmatrix} 7 & 6 & 1 \\ 5 & 8 & 1 \\ 2 & 12 & 0 \\ 9 & 5 & 0 \end{bmatrix} \end{array} = A \qquad \begin{array}{c} \\ C \\ E \\ F \\ O \end{array}\begin{array}{c} \begin{array}{ccc} W & L & T \end{array} \\ \begin{bmatrix} 8 & 6 & 1 \\ 9 & 6 & 0 \\ 5 & 9 & 1 \\ 11 & 4 & 0 \end{bmatrix} \end{array} = B$$

Look carefully at how you answered the questions above. Then look at where those numbers appear in the matrices. How would you add $A + B$?

Take time to think before reading further!

Definition : Matrices of the same dimensions are added by adding corresponding elements.

For instance, a_{ij} corresponds to b_{ij} because they both lie in the ith row and j^{th} column of their respective matrices. Therefore, we would add, $a_{ij} + b_{ij}$ to obtain the $(i,j)^{th}$ element of $A + B$.

$$A+B=\begin{bmatrix} 7 & 6 & 1 \\ 5 & 8 & 1 \\ 2 & 12 & 0 \\ 9 & 5 & 0 \end{bmatrix}+\begin{bmatrix} 8 & 6 & 1 \\ 9 & 6 & 0 \\ 5 & 9 & 1 \\ 11 & 4 & 0 \end{bmatrix}=\begin{bmatrix} 7+8 & 6+6 & 1+1 \\ 5+9 & 8+6 & 1+0 \\ 2+5 & 12+9 & 0+1 \\ 9+11 & 5+4 & 0+0 \end{bmatrix}=\begin{bmatrix} 15 & 12 & 2 \\ 14 & 14 & 1 \\ 7 & 21 & 1 \\ 20 & 9 & 0 \end{bmatrix}$$

Think about the similarities between addition and subtraction. How do you think matrices are subtracted?

Definition : Matrices of the same dimensions are subtracted by subtracting corresponding elements.

Suppose Y represents the wins, losses, and ties for these teams for the entire season (regular season and the playoffs together). Consider the following data

$$\begin{array}{c} \\ C \\ E \\ F \\ O \end{array} \begin{array}{c} \begin{array}{ccc} W & L & T \end{array} \\ \begin{bmatrix} 17 & 13 & 2 \\ 15 & 15 & 1 \\ 7 & 21 & 1 \\ 23 & 9 & 0 \end{bmatrix} \end{array} = Y.$$

How would you find the number of wins, losses, and ties for the playoffs? We would subtract the number of wins, losses, and ties for the regular season from the number of wins, losses, and ties for the entire season.

$$Y-(A+B) = \begin{bmatrix} 17 & 13 & 2 \\ 15 & 15 & 1 \\ 7 & 21 & 1 \\ 23 & 9 & 0 \end{bmatrix} - \begin{bmatrix} 15 & 12 & 2 \\ 14 & 14 & 1 \\ 7 & 21 & 1 \\ 20 & 9 & 0 \end{bmatrix} = \begin{bmatrix} 17-15 & 13-12 & 2-2 \\ 15-14 & 15-14 & 1-1 \\ 7-7 & 21-21 & 1-1 \\ 23-20 & 9-9 & 0-0 \end{bmatrix}$$

$$= \begin{bmatrix} 2 & 1 & 0 \\ 1 & 1 & 0 \\ 0 & 0 & 0 \\ 3 & 0 & 0 \end{bmatrix}$$

Remark : Remember, to add (or subtract) matrices, add (or subtract) corresponding elements. The addition property of zero for real numbers tells us that $r + 0 = 0 + r = r$. There is also an addition property of zero for matrices which states that $A + 0 = 0 + A = A$ where 0 represents the zero matrix of the same dimensions as A.

Definition : A zero matrix is a matrix which has the number 0 for each of its elements.

Remark : We say "*a*" zero matrix instead of "the" zero matrix because for different pairs of dimensions, we have different zero matrices. However for a given pair of dimensions, the zero matrix is unique because zero is unique in the real number system. It is usual to merely say the zero matrix and not refer to its dimensions when no confusion can arise.

Arithmetic

Arithmetic or arithmetics is the oldest and most elementary branch of mathematics, used by almost everyone, for tasks ranging from simple day-to-day counting to advanced science and business calculations. In common usage, the word refers to a branch of (or the forerunner of) mathematics which records elementary properties of certain operations on numbers. Professional mathematicians sometimes use the term *(higher) arithmetic* when referring to number theory, but this should not be confused with elementary arithmetic.

History

The prehistory of arithmetic is limited to a very small number of small artifacts indicating a clear conception of addition and subtraction, the best-known being the Ishango bone from central Africa, dating from somewhere between 20,000 and 18,000 BC.

It is clear that the Babylonians had solid knowledge of almost all aspects of elementary arithmetic by 1800 BC, although historians can only guess at the methods utilized to generate the arithmetical results - as shown, for instance, in the clay tablet Plimpton 322, which appears to be a list of Pythagorean triples, but with no workings to show how the list was originally produced. Likewise, the Egyptian Rhind Mathematical Papyrus (dating from c. 1650 BC, though evidently a copy of an older text from c. 1850 BC) shows evidence of addition, subtraction, multiplication, and division being used within a unit fraction system.

Nicomachus (c. AD 60 - c. AD 120) summarised the philosophical Pythagorean approach to numbers, and their relationships to each other, in his *Introduction to Arithmetic*. At this time, basic arithmetical operations were highly complicated affairs; it was the method known as the "Method of the Indians" that became the arithmetic that we know today. Indian arithmetic was much simpler than Greek arithmetic due to the simplicity of the Indian number system, which had a zero and place-value notation.

The 7th century Syriac bishop Severus Sebokht mentioned this method with admiration, stating however that the Method of the Indians was beyond description. The Arabs learned this new method and called it *hesab*. Fibonacci (also known as Leonardo of Pisa) introduced the "Method of the Indians" to Europe in 1202. In his book *Liber Abaci*, Fibonacci says that, compared with this new method, all other methods had been mistakes. In the Middle Ages, arithmetic was one of the seven liberal arts taught in universities.

Modern algorithms for arithmetic (both for hand and electronic computation) were made possible by the introduction of Arabic numerals and decimal place notation for numbers. Arabic numeral based arithmetic was developed by the great Indian mathematicians Aryabhatta, Brahmagupta and Bhâskara I. Aryabhatta tried different place value notations and Brahmagupta added zero to the Indian number system.

Brahmagupta developed modern multiplication, division, addition and subtraction based on Arabic numerals. Although it is now considered elementary, its simplicity is the culmination of thousands of years of mathematical development.

By contrast, the ancient mathematician Archimedes devoted an entire work, *The Sand Reckoner*, to devising a notation for a certain large integer. The flourishing of algebra in the medieval Islamic world and in Renaissance Europe was an outgrowth of the enormous simplification of computation through decimal notation.

Decimal Arithmetic

Decimal notation constructs all real numbers from the basic digits, the first ten non-negative integers 0,1,2,...,9. A decimal numeral consists of a sequence of these basic digits, with the "denomination" of each digit depending on its *position* with respect to the decimal point: for example, 507.36 denotes 5 hundreds (10^2), plus 0 tens (10^1), plus 7 units (10^0), plus 3 tenths (10^{-1}) plus 6 hundredths (10^{-2}). An essential part of this notation (and a major stumbling block in achieving it) was conceiving of zero as a number comparable to the other basic digits.

Algorism comprises all of the rules of performing arithmetic computations using a decimal system for representing numbers in which numbers written using ten symbols having the values 0 through 9 are combined using a place-value system (positional notation), where each symbol has ten times the weight of the one to its right.

This notation allows the addition of arbitrary numbers by adding the digits in each place, which is accomplished with a 10 x 10 addition table. (A sum of digits which exceeds 9 must have its 10-digit carried to the next place leftward.) One can make a similar algorithm for multiplying arbitrary numbers because the set of denominations

$\{..., 10^2, 10, 1, 10^{-1}, ...\}$ is closed under multiplication. Subtraction and division are achieved by similar, though more complicated algorithms.

Arithmetic Operations

The traditional arithmetic operations are addition, subtraction, multiplication and division, although more advanced operations (such as manipulations of percentages, square root, exponentiation, and logarithmic functions) are also sometimes included in this subject. Arithmetic is performed according to an order of operations. Any set of objects upon which all four operations of arithmetic can be performed (except division by zero), and wherein these four operations obey the usual laws, is called a field.

Addition (+)

Addition is the basic operation of arithmetic. In its simplest form, addition combines two numbers, the *addends* or *terms*, into a single number, the *sum* of the numbers. Adding more than two numbers can be viewed as repeated addition; this procedure is known as summation and includes ways to add infinitely many numbers in an infinite series; repeated addition of the number one is the most basic form of counting.

Addition is commutative and associative so the order in which the terms are added does not matter. The identity element of addition (the additive identity) is 0, that is, adding zero to any number will yield that same number. Also, the inverse element of addition (the additive inverse) is the opposite of any number, that is, adding the opposite of any number to the number itself will yield the additive identity, 0. For example, the opposite of 7 is -7, so 7 + (-7) = 0.

Addition can be given geometrically as follows:

If a and b are the lengths of two sticks, then if we place the sticks one after the other, the length of the stick thus formed will be $a + b$.

Subtraction (–)

Subtraction is the opposite of addition. Subtraction finds the *difference* between two numbers, the *minuend* minus the *subtrahend*. If the minuend is larger than the subtrahend, the difference will be positive; if the minuend is smaller than the subtrahend, the difference will be negative; and if they are equal, the difference will be zero. Subtraction is neither commutative nor associative. For that reason, it is often helpful to look at subtraction as addition of the minuend and the opposite of the subtrahend, that is a " $b = a +$ (eb). When written as a sum, all the properties of addition hold.

*Multiplication (×, ; or *)*

Multiplication is the second basic operation of arithmetic. Multiplication also combines two numbers into a single number, the *product*. The two original numbers are called the *multiplier* and the *multiplicand*, sometimes both simply called *factors*.

Multiplication is best viewed as a scaling operation. If the real numbers are imagined as lying in a line, multiplication by a number, say x, greater than 1 is the same as stretching everything away from zero uniformly, in such a way that the number 1 itself is stretched to where x was. Similarly, multiplying by a number less than 1 can be imagined as squeezing towards zero. (Again, in such a way that 1 goes to the multiplicand.)

Multiplication is commutative and associative; further it is distributive over addition and subtraction.

The multiplicative identity is 1, that is, multiplying any number by 1 will yield that same number. Also, the multiplicative inverse is the reciprocal of any number, that is, multiplying the reciprocal of any number by the number itself will yield the multiplicative identity.

Division (÷ or /)

Division is essentially the opposite of multiplication. Division finds the *quotient* of two numbers, the *dividend* divided by the *divisor*. Any dividend divided by zero is undefined. For positive numbers, if the dividend is larger than the divisor, the quotient will be greater than one, otherwise it will be less than one (a similar rule applies for negative numbers). The quotient multiplied by the divisor always yields the dividend.

Division is neither commutative nor associative. As it is helpful to look at subtraction as addition, it is helpful to look at division as multiplication of the dividend times the reciprocal of the divisor, that is $a \div b = a \times {}^{1}\mathrm{D}_{b}$. When written as a product, it will obey all the properties of multiplication.

Number Theory

The term *arithmetic* is also used to refer to number theory. This includes the properties of integers related to primality, divisibility, and the solution of equations by integers, as well as modern research which is an outgrowth of this study. It is in this context that one runs across the fundamental theorem of arithmetic and arithmetic functions. *A Course in Arithmetic* by Jean-Pierre Serre reflects this usage, as do such phrases as *first order arithmetic* or *arithmetical algebraic geometry*. Number theory is also referred to as *the higher arithmetic*, as in the title of Harold Davenport's book on the subject.

Arithmetic in Education

Primary education in mathematics often places a strong focus on algorithms for the arithmetic of natural numbers, integers, rational numbers (vulgar fractions), and real numbers (using the decimal place-value system). This study is sometimes known as algorism.

The difficulty and unmotivated appearance of these algorithms has long led educators to question this curriculum, advocating the early teaching of more central and intuitive mathematical ideas. One notable movement in this direction was the New Math of the 1960s and 1970s, which attempted to teach arithmetic in the spirit of axiomatic development from set theory, an echo of the prevailing trend in higher mathematics.

Since the introduction of the electronic calculator, which can perform the algorithms far more efficiently than humans, an influential school of educators has argued that mechanical mastery of the standard arithmetic algorithms is no longer necessary.

In their view, the first years of school mathematics could be more profitably spent on understanding higher-level ideas about what numbers are used for and relationships among number, quantity, measurement, and so on. However, most research mathematicians still consider mastery of the manual algorithms to be a necessary foundation for the study of algebra and computer science. This controversy was central to the "math wars" over California's primary school curriculum in the 1990s, and continues today.

Many mathematics texts for K–12 instruction were developed, funded by grants from the United States National Science Foundation based on standards created by the National Council of Teachers of Mathematics and given high ratings by United States Department of Education, though condemned by many mathematicians. Some widely adopted texts such as Investigations in Number, Data, and Space, developed by the education research organization TERC, were based on the spirit of research papers which found that instruction of basic arithmetic was harmful to mathematical understanding.

Rather than teaching any traditional method of arithmetic, teachers are instructed instead to guide students to invent their own (some critics claim inefficient) methods, using for example such techniques as skip counting, and the heavy use of manipulatives, scissors and paste, and even singing, rather than multiplication tables or long division.

Although such texts were designed to be complete curricula, in the face of intense protest and criticism, many school districts have chosen to circumvent the intent of such radical approaches by supplementing with traditional texts. Other districts have since adopted traditional mathematics texts, and discarded such reform-based approaches as misguided failures.

Modular Arithmetic

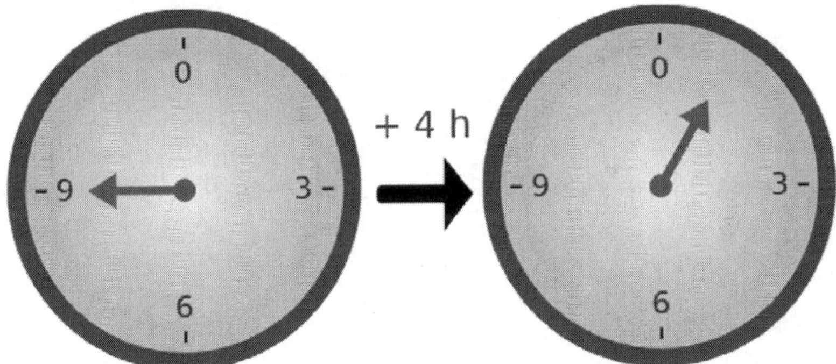

Figure: *Time-keeping on this clock uses arithmetic modulo 12.*

A familiar use of modular arithmetic is its use in the 12-hour clock, in which the day is divided into two 12 hour periods. If the time is 7:00 now, then 8 hours later it will be 3:00. Usual addition would suggest that the later time should be $7 + 8 = 15$, but this is not the answer because clock time "wraps around" every 12 hours; there is no "15 o'clock".

Likewise, if the clock starts at 12:00 (noon) and 21 hours elapse, then the time will be 9:00 the next day, rather than 33:00. Since the hour number starts over when it reaches 12, this is arithmetic *modulo* 12. 12 is congruent not only to 12 itself, but also to 0, so the time called "12:00" could also be called "0:00", since $0 = 12 \bmod 12$.

The Congruence Relation

Modular arithmetic can be handled mathematically by introducing a congruence relation on the integers that is compatible with the operations of the ring of integers: addition, subtraction, and multiplication. For a positive integer n, two integers a and b are said to be congruent modulo n, written:

$$a \equiv b \pmod{n},$$

if their difference a " b is an integer multiple of n (or n divides a " b). The number n is called the modulus of the congruence.

For example,

$$38 \equiv 14 \pmod{12}$$

because 38 " 14 = 24, which is a multiple of 12.

The same rule holds for negative values:

$$-8 \equiv 7 \pmod{5}.$$

$$2 \equiv -3 \pmod{5}.$$

$$-3 \equiv -8 \pmod{5}.$$

Equivalently, $a \equiv b \pmod{n}$ can also be thought of as asserting that the remainders of the division of both a and b by n are the same. For instance:

$$38 \equiv 14 \pmod{12}$$

because both 38 and 14 have the same remainder 2 when divided by 12. It is also the case that $38 - 14 = 24$ is an integer multiple of 12, which agrees with the prior definition of the congruence relation.

A remark on the notation: Because it is common to consider several congruence relations for different moduli at the same time, the modulus is incorporated in the notation. In spite of the ternary notation, the congruence relation for a given modulus isbinary. This would have been clearer if the notation a aH$_n$ b had been used, instead of the common traditional notation.

The properties that make this relation a congruence relation (respecting addition, subtraction, and multiplication) are the following.

If

$$a_1 \equiv b_1 \pmod{n}$$

and

$$a_2 \equiv b_2 \pmod{n},$$

then:

- $a_1 + a_2 \equiv b_1 + b_2 \pmod{n}$
- $a_1 - a_2 \equiv b_1 - b_2 \pmod{n}$

It should be noted that the above two properties would still hold if the theory were expanded to include all real numbers, that is if a_1, a_2, b_1, b_2, n were not necessarily all integers. The next property, however, would fail if these variables were not all integers:

- $a_1 a_2 \equiv b_1 b_2 \pmod{n}.$

The Ring of Congruence Classes

Like any congruence relation, congruence modulo n is an equivalence relation, and the equivalence class of the integer a, denoted by $\bar{a}_n$, is the set $\{\ldots, a-2n, a-n, a, a+n, a+2n, \ldots\}$.

This set, consisting of the integers congruent to a modulo n, is called the congruence class or residue class of a modulo n. Another notation for this congruence class, which requires that in the context the modulus is known, is $[a]$.

Integers Modulo n

The set of all congruence classes of the integers for a modulus n is called the set of integers modulo n, and is denoted $\mathbb{Z}/n\mathbb{Z}$, $\mathbb{Z}/n$, or $\mathbb{Z}_n$. The notation $\mathbb{Z}_n$ is, however, not recommended because it can be confused with the set of n-adic integers. The set is defined as follows.

$$\mathbb{Z}/n\mathbb{Z} = \{\bar{a}_n \mid a \in \mathbb{Z}\}.$$

When $n \neq 0$, $\mathbb{Z}/n\mathbb{Z}$ has n elements, and can be written as:

$$\mathbb{Z}/n\mathbb{Z} = \left\{\bar{0}_n, \bar{1}_n, \bar{2}_n, \ldots, \overline{n-1}_n\right\}.$$

When n = 0, $\mathbb{Z}/n\mathbb{Z}$ does not have zero elements; rather, it is isomorphic to $\mathbb{Z}$, since $\bar{a}_0 = \{a\}$.

We can define addition, subtraction, and multiplication on $\mathbb{Z}/n\mathbb{Z}$ by the following rules:

- $\bar{a}_n + \bar{b}_n = \overline{(a+b)}_n$

- $\overline{a}_n - \overline{b}_n = \overline{(a-b)}_n$
- $\overline{a}_n \overline{b}_n = \overline{(ab)}_n$.

The verification that this is a proper definition uses the properties given before.

In this way, $\mathbb{Z}/n\mathbb{Z}$ becomes a commutative ring. For example, in the ring $\mathbb{Z}/24\mathbb{Z}$, we have $\overline{12}_{24} + \overline{21}_{24} = \overline{9}_{24}$ as in the arithmetic for the 24-hour clock.

The notation $\mathbb{Z}/n\mathbb{Z}$ is used, because it is the factor ring of $\mathbb{Z}$ by the ideal $n\mathbb{Z}$ containing all integers divisible by n, where $0\mathbb{Z}$ is the singleton set $\{0\}$. Thus $\mathbb{Z}/n\mathbb{Z}$ is a field when $n\mathbb{Z}$ is a maximal ideal, that is, when n is prime.

In terms of groups, the residue class $\overline{a}_n$ is the coset of a in the quotient group $\mathbb{Z}/n\mathbb{Z}$, a cyclic group. The set $\mathbb{Z}/n\mathbb{Z}$ has a number of important mathematical properties that are foundational to various branches of mathematics.

Rather than excluding the special case n = 0, it is more useful to include $\mathbb{Z}/0\mathbb{Z}$ (which, as mentioned before, is isomorphic to the ring $\mathbb{Z}$ of integers), for example when discussing the characteristic of a ring.

When n is prime, the set of integers modulo n form a finite field.

Remainders

The notion of modular arithmetic is related to that of the remainder in division. The operation of finding the remainder is sometimes referred to as the modulo operation and we may see 2 = 14 (mod 12).

The difference is in the use of congruency, indicated by "$\equiv$", and equality indicated by "=". Equality implies specifically the "common residue", the least non-negative member of an equivalence class. When working with modular arithmetic, each equivalence class is usually represented by its common residue, for example 38 ≡ 2 (mod 12) which can be found using long division. It follows that, while it is correct to say 38 ≡ 14 (mod 12), and 2 ≡ 14 (mod 12), it is incorrect to say 38 = 14 (mod 12) (with "=" rather than "≡").

The difference is clearest when dividing a negative number, since in that case remainders are negative. Hence to express the remainder we would have to write –5 ≡ –17 (mod 12), rather than 7 = –17 (mod 12), since "=" can only be used with the (positive) common residue.

In computer science, it is the remainder operator that is usually indicated by either "%" (e.g. in C, Java, Javascript, and Python) or "mod" (e.g. in SQL, Visual Basic), with exceptions (e.g. Excel).

These operators are commonly pronounced as "mod", but it is specifically a remainder that is computed (since a negative number will be returned if the first argument is negative).

The function *modulo* instead of *mod*, like 38 ≡ 14 (modulo 12) is sometimes used to indicate the common residue rather than a remainder (e.g. in Ruby).

Parentheses are sometimes dropped from the expression, e.g. 38 ≡ 14 mod 12 or 2 = 14 mod 12, or placed around the divisor e.g. 38 ≡ 14 mod (12).

Notation such as 38(mod 12) has also been observed, but is ambiguous without contextual clarification.

Functional Representation of the Remainder

Computation with modular arithmetic can be implemented using other functions.

One such functional representation uses the floor function. If $a \equiv b \pmod{n}$, $a \geq 0$, and $0 \leq b < n$, then there exists an integer $k \geq 0$ such that $a = kn + b$. The remainder b can be written

$$b = a - \left[\frac{a}{n}\right] \times n,$$

where $\left[\frac{a}{n}\right]$ is the integer (whole) part of $\frac{a}{n}$. If instead $-n \leq b < 0$, then

$$b = a - \left[\frac{a}{n}\right] \times n - n.$$

Another functional representation uses sine and arcsine, taking advantage of the fact that arcsine is multivalued. Let

$$g(x) = \frac{2n}{\pi} \arcsin\left(\sin\left(\frac{\pi x}{2n}\right)\right).$$

Then

$$g(x) = x \pmod{n}$$

for

$$0 \leq \pi x/2n < \pi/2$$

or

$$\pi \leq \pi x/2n < 3\pi/2,$$

and

$$g(x) = -x \text{ (mod } n)$$

for

$$\pi/2 \leq \pi x/2n < \pi \text{ or } 3\pi/2 \leq \pi x/2n < 2\pi.$$

Applications

Modular arithmetic is referenced in number theory, group theory, ring theory, knot theory, abstract algebra, cryptography, computer science, chemistry and the visual and musical arts.

It is one of the foundations of number theory, touching on almost every aspect of its study, and provides key examples for group theory, ring theory and abstract algebra.

In cryptography, modular arithmetic directly underpins public key systems such as RSA and Diffie-Hellman, as well as providing finite fields which underlie elliptic curves, and is used in a variety of symmetric key algorithms including AES, IDEA, and RC4.

In computer science, modular arithmetic is often applied in bitwise operations and other operations involving fixed-width, cyclic data structures. The modulo operation, as implemented in many programming languages and calculators, is an application of modular arithmetic that is often used in this context.

In chemistry, the last digit of the CAS registry number (a number which is unique for each chemical compound) is a check digit, which is calculated by taking the last digit of the first two parts of the CAS registry number times 1, the next digit times 2, the next digit times 3 etc., adding all these up and computing the sum modulo 10.

In music, arithmetic modulo 12 is used in the consideration of the system of twelve-tone equal temperament, where octave and enharmonic equivalency occurs (that is, pitches in a 16"2 or 26"1 ratio are equivalent, and C-sharp is considered the same as D-flat).

The method of casting out nines offers a quick check of decimal arithmetic computations performed by hand. It is based on modular arithmetic modulo 9, and specifically on the crucial property that 10 ≡ 1 (mod 9). More generally, modular arithmetic also has application in disciplines such as law, economics, and other areas of the social sciences, where proportional division and allocation of resources plays a central part of the analysis.

All your congruence are belong to us.

Computational Complexity

Since modular arithmetic has such a wide range of applications, it is important to know how hard it is to solve a system of congruences. A linear system of congruences can be solved in polynomial time with a form of Gaussian elimination, for details see Linear congruence theorem. Algorithms, such as Montgomery reduction, also exist to allow simple arithmetic operations, such as multiplication and exponentiation modulo n, to be performed efficiently on large numbers.

System of Linear Equations

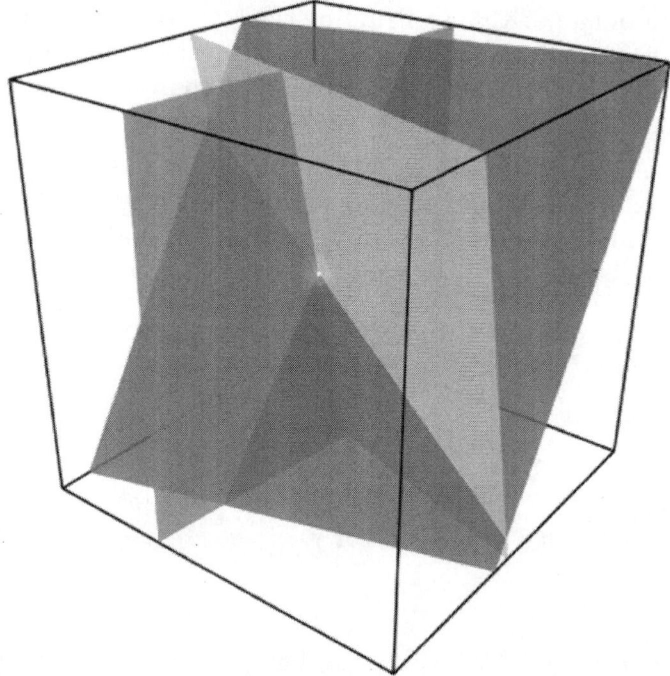

Figure: *A linear system in three variables determines a collection of planes. The intersection point is the solution.*

In mathematics, a system of linear equations (or linear system) is a collection of linear equations involving the same set of variables. For example,

$$3x + 2y - z = 1$$

$$2x - 2y + 4z = -2$$
$$-x + \tfrac{1}{2}y - z = 0$$

is a system of three equations in the three variables x, y, z. A solution to a linear system is an assignment of numbers to the variables such that all the equations are simultaneously satisfied. A solution to the system above is given by

$$x = 1$$
$$y = -2$$
$$z = -2$$

since it makes all three equations valid.

In mathematics, the theory of linear systems is a branch of linear algebra, a subject which is fundamental to modern mathematics. Computational algorithms for finding the solutions are an important part of numerical linear algebra, and such methods play a prominent role in engineering, physics, chemistry, computer science, and economics.

A system of non-linear equations can often be approximated by a linear system, a helpful technique when making a mathematical model or computer simulation of a relatively complex system.

Elementary Example

The simplest kind of linear system involves two equations and two variables:

$$2x + 3y = 6$$
$$4x + 9y = 15.$$

One method for solving such a system is as follows. First, solve the top equation for x in terms of y:

$$x = 3 - \frac{3}{2}y.$$

Now substitute this expression for x into the bottom equation:

$$4\left(3 - \frac{3}{2}y\right) + 9y = 15.$$

This results in a single equation involving only the variable y. Solving gives $y = 1$, and substituting this back into the equation for x yields $x = 3/2$. This method generalizes to systems with additional variables.

General Form

A general system of m linear equations with n unknowns can be written as

$$\begin{aligned} a_{11}x_1 + a_{12}x_2 + \dots + a_{1n}x_n &= b_1 \\ a_{21}x_1 + a_{22}x_2 + \dots + a_{2n}x_n &= b_2 \\ \vdots \quad \vdots \quad & \quad \vdots \quad \vdots \\ a_{m1}x_1 + a_{m2}x_2 + \dots + a_{mn}x_n &= b_{m.} \end{aligned}$$

Here $x_1, x_2, \dots, x_n$ are the unknowns, $a_{11}, a_{12}, \dots, a_{mn}$ are the coefficients of the system, and $b_1, b_2, \dots, b_m$ are the constant terms.

Often the coefficients and unknowns are real or complex numbers, but integers and rational numbers are also seen, as are polynomials and elements of an abstract algebraic structure.

Vector Equation

One extremely helpful view is that each unknown is a weight for a column vector in a linear combination.

$$x_1 \begin{bmatrix} a_{11} \\ a_{21} \\ \vdots \\ a_{m1} \end{bmatrix} + x_2 \begin{bmatrix} a_{12} \\ a_{22} \\ \vdots \\ a_{m2} \end{bmatrix} + \cdots + x_n \begin{bmatrix} a_{1n} \\ a_{2n} \\ \vdots \\ a_{mn} \end{bmatrix} = \begin{bmatrix} b_1 \\ b_2 \\ \vdots \\ b_m \end{bmatrix}$$

This allows all the language and theory of *vector spaces* (or more generally, *modules*) to be brought to bear. For example, the collection of all possible linear combinations of the vectors on the left-hand side is called their *span*, and the equations have a solution just when the right-hand vector is within that span.

If every vector within that span has exactly one expression as a linear combination of the given left-hand vectors, then any solution is unique.

In any event, the span has a *basis* of linearly independent vectors that do guarantee exactly one expression; and the number of vectors in that basis (its *dimension*) cannot be larger than m or n, but it can be smaller.

This is important because if we have m independent vectors a solution is guaranteed regardless of the right-hand side, and otherwise not guaranteed.

Matrix Equation

The vector equation is equivalent to a matrix equation of the form

$$Ax = b$$

where A is an $m \times n$ matrix, x is a column vector with n entries, and b is a column vector with m entries.

$$A = \begin{bmatrix} a_{11} & a_{12} & \cdots & a_{1n} \\ a_{21} & a_{22} & \cdots & a_{2n} \\ \vdots & \vdots & \ddots & \vdots \\ a_{m1} & a_{m2} & \cdots & a_{mn} \end{bmatrix}, \quad x = \begin{bmatrix} x_1 \\ x_2 \\ \vdots \\ x_n \end{bmatrix}, \quad b = \begin{bmatrix} b_1 \\ b_2 \\ \vdots \\ b_m \end{bmatrix}$$

The number of vectors in a basis for the span is now expressed as the *rank* of the matrix.

Solution Set

The solution set for the equations $x - y = -1$ and $3x + y = 9$ is the single point (2, 3).

A solution of a linear system is an assignment of values to the variables $x_1, x_2, \ldots, x_n$ such that each of the equations is satisfied. The set of all possible solutions is called the solution set.

A linear system may behave in any one of three possible ways:

1. The system has *infinitely many solutions.*
2. The system has a single *unique solution.*
3. The system has no *solution.*

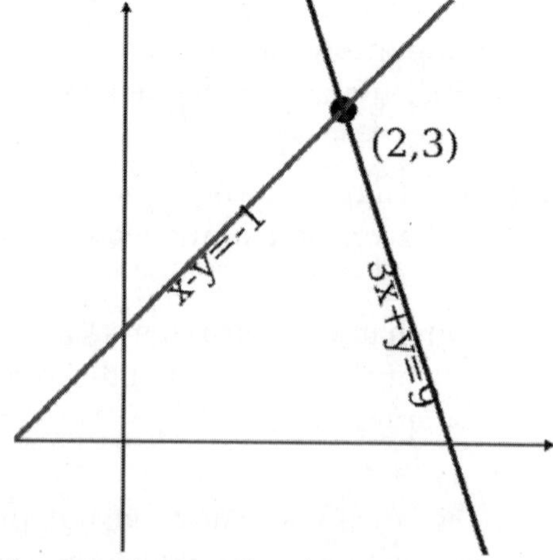

Figure: *The solution set for the equations x - y = -1 and 3x + y = 9 is the single point (2, 3).*

Geometric Interpretation

For a system involving two variables (x and y), each linear equation determines a line on the xy-plane. Because a solution to a linear system must satisfy all of the equations, the solution set is the intersection of these lines, and is hence either a line, a single point, or the empty set.

For three variables, each linear equation determines a plane in three-dimensional space, and the solution set is the intersection of these planes. Thus the solution set may be a plane, a line, a single point, or the empty set. For n variables, each linear equations determines a hyperplane in n-dimensional space. The solution set is the intersection of these hyperplanes, which may be a flat of any dimension.

General behaviour

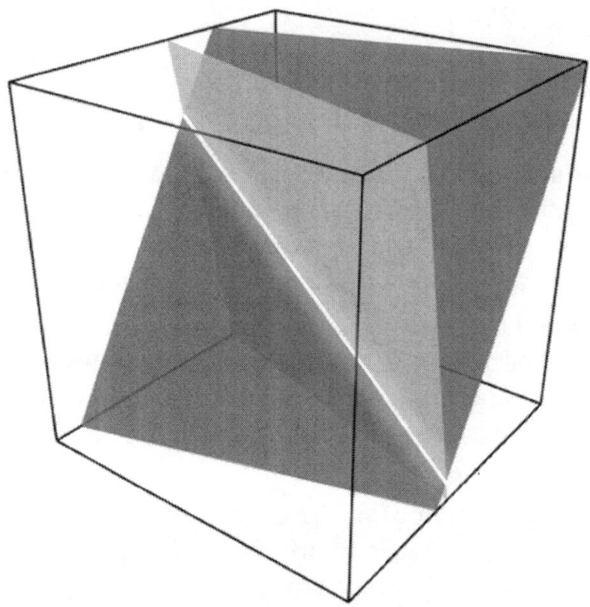

Figure: *The solution set for two equations in three variables is usually a line.*

In general, the behaviour of a linear system is determined by the relationship between the number of equations and the number of unknowns:

1. Usually, a system with fewer equations than unknowns has infinitely many solutions. Such a system is also known as an underdetermined system.

2. Usually, a system with the same number of equations and unknowns has a single unique solution.
3. Usually, a system with more equations than unknowns has no solution.

In the first case, the dimension of the solution set is usually equal to $n - m$, where n is the number of variables and m is the number of equations.

The following pictures illustrate this trichotomy in the case of two variables:

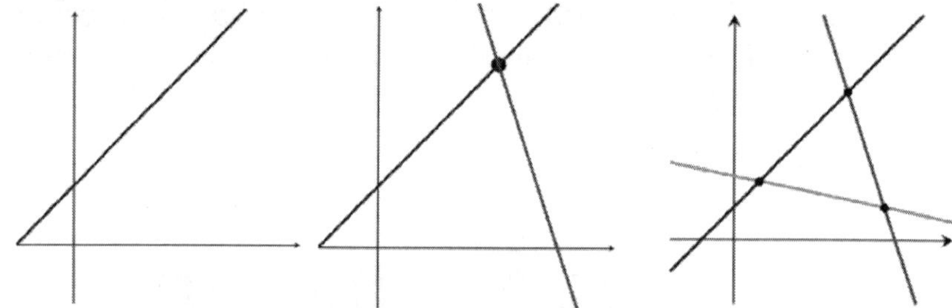

Figure: *One equation* ***Figure:*** *Two equations* ***Figure:*** *Three equations*

The first system has infinitely many solutions, namely all of the points on the blue line. The second system has a single unique solution, namely the intersection of the two lines. The third system has no solutions, since the three lines share no common point.

Keep in mind that the pictures above show only the most common case. It is possible for a system of two equations and two unknowns to have no solution (if the two lines are parallel), or for a system of three equations and two unknowns to be solvable (if the three lines intersect at a single point).

In general, a system of linear equations may behave differently than expected if the equations are linearly dependent, or if two or more of the equations are inconsistent.

Properties

Consistency

The equations of a linear system are consistent if they possess a common solution, and inconsistent otherwise. When the equations are inconsistent, it is possible to derive a contradiction from the equations, such as the statement that $0 = 1$.

For Example, the Equations:

$$3x + 2y = 6 \text{ and } 3x + 2y = 12$$

are inconsistent. In attempting to find a solution, we tacitly assume that there is a solution; that is, we assume that the value of x in the first equation must be the same as the value of x in the second equation (the same is assumed to simultaneously be true for the value of y in both equations).

Applying the substitution property (for 3x+2y) yields the equation 6 = 12, which is a false statement. This therefore contradicts our assumption that the system had a solution and we conclude that our assumption was false; that is, the system in fact has no solution. The graphs of these equations on the xy-plane are a pair of parallel lines.

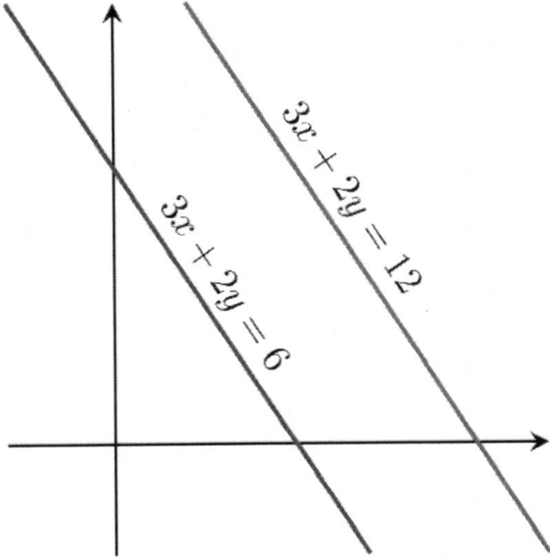

Figure: *The equations 3x + 2y = 6 and 3x + 2y = 12 are inconsistent.*

It is possible for three linear equations to be inconsistent, even though any two of the equations are consistent together. For example, the equations

$$x + y = 1$$
$$2x + y = 1$$
$$3x + 2y = 3$$

are inconsistent. Adding the first two equations together gives $3x + 2y = 2$, which can be subtracted from the third equation to yield 0 = 1. Note that any two of these equations have a common solution. The same phenomenon can occur for any number of equations. In general, inconsistencies occur if the left-hand sides of the equations in a system are linearly dependent, and the constant terms do not satisfy the dependence relation. A system of equations whose left-hand sides are linearly independent is always consistent.

Independence

The equations of a linear system are independent if none of the equations can be derived algebraically from the others. When the equations are independent, each equation contains new information about the variables, and removing any of the equations increases the size of the solution set.

For linear equations, logical independence is the same as linear independence. The equations $x - 2y = -1$, $3x + 5y = 8$, and $4x + 3y = 7$ are not linearly independent.

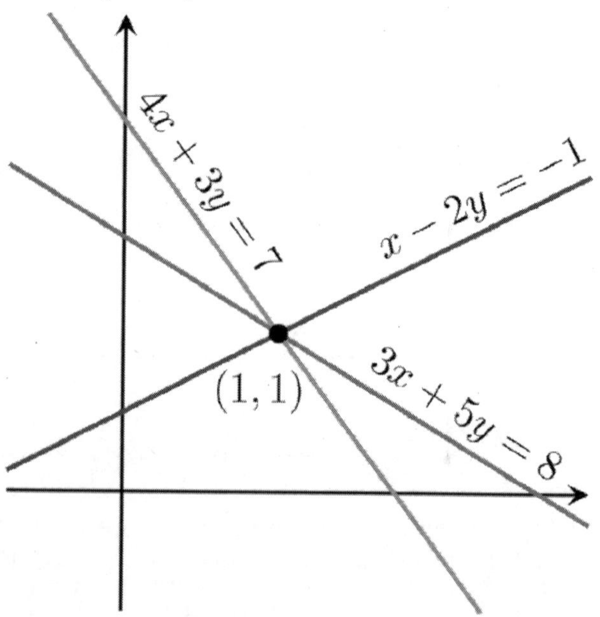

Figure: *The equations x - 2y = -1, 3x + 5y = 8, and 4x + 3y = 7 are linearly dependent.*

For example, the equations

$$3x+2y=6 \text{ and } 6x+4y=12$$

are not independent- they are the same equation when scaled by a factor of two, and they would produce identical graphs. This is an example of equivalence in a system of linear equations.

For a more complicated example, the equations

$$x - 2y = -1$$
$$3x + 5y = 8$$
$$4x + 3y = 7$$

are not independent, because the third equation is the sum of the other two. Indeed, any one of these equations can be derived from the other two, and any one of the equations can be removed without affecting the solution set. The graphs of these equations are three lines that intersect at a single point.

Equivalence

Two linear systems using the same set of variables are equivalent if each of the equations in the second system can be derived algebraically from the equations in the first system, and vice-versa. Equivalent systems convey precisely the same information about the values of the variables. In particular, two linear systems are equivalent if and only if they have the same solution set.

Solving a Linear System

There are several algorithms for solving a system of linear equations.

Describing the Solution

When the solution set is finite, it is usually described in set notation. For example, the solution set 2, 3, and 4 would be written: {2,3,4}

It can be difficult to describe a set with infinite solutions. Typically, some of the variables are designated as free (or independent, or as parameters), meaning that they are allowed to take any value, while the remaining variables are dependent on the values of the free variables.

For example, consider the following system:

$$x + 3y - 2z = 5$$
$$3x + 5y + 6z = 7$$

The solution set to this system can be described by the following equations:

$$x = -7z - 1 \text{ and } y = 3z + 2.$$

Here z is the free variable, while x and y are dependent on z. Any point in the solution set can be obtained by first choosing a value for z, and then computing the corresponding values for x and y.

Each free variable gives the solution space one degree of freedom, the number of which is equal to the dimension of the solution set.

For example, the solution set for the above equation is a line, since a point in the solution set can be chosen by specifying the value of the parameter z.

An infinite solution of higher order may describe a plane, or higher dimensional set.

Different choices for the free variables may lead to different descriptions of the same solution set.

For example, the solution to the above equations can alternatively be described as follows:

$$y = -\frac{3}{7}x + \frac{11}{7} \text{ and } z = -\frac{1}{7}x - \frac{1}{7}.$$

Here x is the free variable, and y and z are dependent.

Elimination of Variables

The simplest method for solving a system of linear equations is to repeatedly eliminate variables. This method can be described as follows:

1. In the first equation, solve for the one of the variables in terms of the others.
2. Plug this expression into the remaining equations. This yields a system of equations with one fewer equation and one fewer unknown.
3. Continue until you have reduced the system to a single linear equation.
4. Solve this equation, and then back-substitute until the entire solution is found.

For example, consider the following system:

$$x + 3y - 2z = 5$$

$$3x + 5y + 6z = 7$$

$$2x + 4y + 3z = 8$$

Solving the first equation for x gives $x = 5 + 2z - 3y$, and plugging this into the second and third equation yields

$$-4y + 12z = -8$$
$$-2y + 7z = -2$$

Solving the first of these equations for y yields $y = 2 + 3z$, and plugging this into the third equation yields $z = 2$. We now have:

$$x = 5 + 2z - 3y$$
$$y = 2 + 3z$$
$$z = 2$$

Substituting $z = 2$ into the second equation gives $y = 8$, and substituting $z = 2$ and $y = 8$ into the first equation yields $x = -15$. Therefore, the solution set is the single point $(x, y, z) = (-15, 8, 2)$.

Row Reduction

In row reduction, the linear system is represented as an augmented matrix:

$$\left[\begin{array}{ccc|c} 1 & 3 & -2 & 5 \\ 3 & 5 & 6 & 7 \\ 2 & 4 & 3 & 8 \end{array}\right].$$

This matrix is then modified using elementary row operations until it reaches reduced row echelon form.

There are three types of elementary row operations:

Type 1: Swap the positions of two rows.

Type 2: Multiply a row by a nonzero scalar.

Type 3: Add to one row a scalar multiple of another.

Because these operations are reversible, the augmented matrix produced always represents a linear system that is equivalent to the original.

There are several specific algorithms to row-reduce an augmented matrix, the simplest of which are Gaussian elimination and Gauss-Jordan elimination.

The following computation shows Gaussian elimination applied to the matrix above:

$$\begin{bmatrix} 1 & 3 & -2 & 5 \\ 3 & 5 & 6 & 7 \\ 2 & 4 & 3 & 8 \end{bmatrix} \sim \begin{bmatrix} 1 & 3 & -2 & 5 \\ 0 & -4 & 12 & -8 \\ 2 & 4 & 3 & 8 \end{bmatrix} \sim \begin{bmatrix} 1 & 3 & -2 & 5 \\ 0 & -4 & 12 & -8 \\ 0 & -2 & 7 & -2 \end{bmatrix} \sim \begin{bmatrix} 1 & 3 & -2 & 5 \\ 0 & 1 & -3 & 2 \\ 0 & -2 & 7 & -2 \end{bmatrix}$$

$$\sim \begin{bmatrix} 1 & 3 & -2 & 5 \\ 0 & 1 & -3 & 2 \\ 0 & 0 & 1 & 2 \end{bmatrix} \sim \begin{bmatrix} 1 & 3 & -2 & 5 \\ 0 & 1 & 0 & 8 \\ 0 & 0 & 1 & 2 \end{bmatrix} \sim \begin{bmatrix} 1 & 3 & 0 & 9 \\ 0 & 1 & 0 & 8 \\ 0 & 0 & 1 & 2 \end{bmatrix} \sim \begin{bmatrix} 1 & 0 & 0 & -15 \\ 0 & 1 & 0 & 8 \\ 0 & 0 & 1 & 2 \end{bmatrix}.$$

The last matrix is in reduced row echelon form, and represents the system $x = -15$, $y = 8$, $z = 2$. A comparison with the example in the previous section on the algebraic elimination of variables shows that these two methods are in fact the same; the difference lies in how the computations are written down.

Cramer's Rule

Cramer's rule is an explicit formula for the solution of a system of linear equations, with each variable given by a quotient of two determinants. For example, the solution to the system

$$\begin{aligned} x &+ 3y - 2z = 5 \\ 3x &+ 5y + 6z = 7 \\ 2x &+ 4y + 3z = 8 \end{aligned}$$

is given by

$$x = \frac{\begin{vmatrix} 5 & 3 & -2 \\ 7 & 5 & 6 \\ 8 & 4 & 3 \end{vmatrix}}{\begin{vmatrix} 1 & 3 & -2 \\ 3 & 5 & 6 \\ 2 & 4 & 3 \end{vmatrix}}, \quad y = \frac{\begin{vmatrix} 1 & 5 & -2 \\ 3 & 7 & 6 \\ 2 & 8 & 3 \end{vmatrix}}{\begin{vmatrix} 1 & 3 & -2 \\ 3 & 5 & 6 \\ 2 & 4 & 3 \end{vmatrix}}, \quad z = \frac{\begin{vmatrix} 1 & 3 & 5 \\ 3 & 5 & 7 \\ 2 & 4 & 8 \end{vmatrix}}{\begin{vmatrix} 1 & 3 & -2 \\ 3 & 5 & 6 \\ 2 & 4 & 3 \end{vmatrix}}.$$

For each variable, the denominator is the determinant of the matrix of coefficients, while the numerator is the determinant of a matrix in which one column has been replaced by the vector of constant terms.

Though Cramer's rule is important theoretically, it has little practical value for large matrices, since the computation of large determinants is somewhat cumbersome. (Indeed, large determinants are most easily computed using row reduction.)

Further, Cramer's rule has very poor numerical properties, making it unsuitable for solving even small systems reliably, unless the operations are performed in rational arithmetic with unbounded precision.

Other Methods

While systems of three or four equations can be readily solved by hand, computers are often used for larger systems. The standard algorithm for solving a system of linear equations is based on Gaussian elimination with some modifications.

Firstly, it is essential to avoid division by small numbers, which may lead to inaccurate results. This can be done by reordering the equations if necessary, a process known as *pivoting*.

Secondly, the algorithm does not exactly do Gaussian elimination, but it computes the LU decomposition of the matrix A.

This is mostly an organizational tool, but it is much quicker if one has to solve several systems with the same matrix A but different vectors b.

If the matrix A has some special structure, this can be exploited to obtain faster or more accurate algorithms.

For instance, systems with a symmetric positive definite can be solved twice as fast with the Cholesky decomposition. Levinson recursion is a fast method for Toeplitz matrices. Special methods exist also for matrices with many zero elements (so-called sparse matrices), which appear often in applications.

A completely different approach is often taken for very large systems, which would otherwise take too much time or memory. The idea is to start with an initial approximation to the solution (which does not have to be accurate at all), and to change this approximation in several steps to bring it closer to the true solution. Once the approximation is sufficiently accurate, this is taken to be the solution to the system. This leads to the class of iterative methods.

Homogeneous Systems

A system of linear equations is homogeneous if all of the constant terms are zero:

$$\begin{array}{ccccccc} a_{11}x_1 & + & a_{12}x_2 & +\cdots+ & a_{1n}x_n & = & 0 \\ a_{21}x_1 & + & a_{22}x_2 & +\cdots+ & a_{2n}x_n & = & 0 \\ \vdots & & \vdots & & \vdots & & \vdots \\ a_{m1}x_1 & + & a_{m2}x_2 & +\cdots+ & a_{mn}x_n & = & 0. \end{array}$$

A homogeneous system is equivalent to a matrix equation of the form

$$Ax = 0$$

where A is an $m \times n$ matrix, x is a column vector with n entries, and 0 is the zero vector with m entries.

Solution Set

Every homogeneous system has at least one solution, known as the zero solution (or trivial solution), which is obtained by assigning the value of zero to each of the variables. The solution set has the following additional properties:

1. If u and v are two vectors representing solutions to a homogeneous system, then the vector sum u + v is also a solution to the system.
2. If u is a vector representing a solution to a homogeneous system, and r is any scalar, then ru is also a solution to the system.

These are exactly the properties required for the solution set to be a linear subspace of R^n.

In particular, the solution set to a homogeneous system is the same as the null space of the corresponding matrix A.

Relation to Nonhomogeneous Systems

There is a close relationship between the solutions to a linear system and the solutions to the corresponding homogeneous system:

$$Ax = b \text{ and } Ax = 0.$$

Specifically, if p is any specific solution to the linear system Ax = b, then the entire solution set can be described as

$$\{p + v : v \text{ is any solution to } Ax = 0\}.$$

Geometrically, this says that the solution set for Ax = b is a translation of the solution set for Ax = 0. Specifically, the flat for the first system can be obtained by translating the linear subspace for the homogeneous system by the vector p.

This reasoning only applies if the system Ax = b has at least one solution. This occurs if and only if the vector b lies in the image of the linear transformation A.

Matrix Solution of Linear Systems

A matrix (plural: matrices) is a rectangular array of numbers arranged in rows and columns and enclosed in brackets. Each number in the matrix is called an element of the matrix.

Each of the following is an example of a matrix:

$$\begin{bmatrix} 5 & -1 & 0 \\ -2 & 6 & 9 \end{bmatrix} \begin{bmatrix} 4 & -3 \\ -7 & 10 \end{bmatrix}$$

We are going to look at a technique for solving linear systems of equations using matrices. A matrix derived from a linear system of equations, each in standard form, is called the augmented matrix of the system. The augmented matrix for a linear system is a matrix with the system's variables and equal signs eliminated. For example:

System

$$x - 2y - z = 2$$

$$2x - y = 4$$

$$-x + y - 2z = -4$$

Augmented Matrix

$$\begin{bmatrix} 1 & -2 & -1 & 2 \\ 2 & -1 & 0 & 4 \\ -1 & 1 & -2 & -4 \end{bmatrix}$$

Each row of the augmented matrix represents one equation of the system. In each row, the coefficient of x is in the first column, the coefficient of y is in the second column, the coefficient of z is in the third column, and the constant is in the fourth column.

Notice that if there is a missing term, as the "z" term in the second equation, a zero represents its coefficient in the augmented matrix.

Matrix Row Transformations

For any augmented matrix of a system of linear equations, the following row transformations will result in the matrix of an equivalent system:

1. Any two rows may be interchanged.
2. The elements of any row may be multiplied (or divided) by the same nonzero number.
3. Any row may be changed by adding to its elements a multiple of the corresponding elements of another row.

To solve a system of equations using the augmented matrix, we will use matrix row transformations to convert the augmented matrix into triangular form. A matrix is in triangular form if there are 1's down the main diagonal and 0's below the diagonal.

The following matrix is in triangular form:

$$\begin{bmatrix} 1 & A & B & C \\ 0 & 1 & D & E \\ 0 & 0 & 1 & F \end{bmatrix}$$

The system of equations represented by this matrix is:

$$x + Ay + Bz = C$$

$$y + Dz = E$$

$$z = F$$

The system is easily solved from this form.

To solve a linear system using matrix row transformations, we begin with the augmented matrix.

We then use row transformations to obtain an equivalent matrix with 1's down the diagonal and 0's below the diagonal.

We will proceed column by column from left to right.

In each column, we will start by obtaining a 1 in the diagonal position; then obtain 0's below the 1.

This method is called the Gauss-Jordan method.

Example. Solve:

$$x + y - z = -2$$

$$2x - y + z = 5$$

$$-x + 2y + 2z = 1$$

Solution.

$$\begin{bmatrix} 1 & 1 & -1 & -2 \\ 2 & -1 & 1 & 5 \\ -1 & 2 & 2 & 1 \end{bmatrix}$$

We begin by writing the system as an augmented matrix

We already have a 1 in the diagonal position of first column. Now we want 0's below the 1.

The first 0 can be obtained by multiplying row 1 by -2 and adding the results to row 2 :

$$\begin{bmatrix} 1 & 1 & -1 & -2 \\ 0 & -3 & 3 & 9 \\ -1 & 2 & 2 & 1 \end{bmatrix}$$

Row 1 is unchanged:

-2 times row 1 is added to row 2

Row 3 is unchanged

The second 0 can be obtained by adding row 1 to row 3:

$$\begin{bmatrix} 1 & 1 & -1 & -2 \\ 0 & -3 & 3 & 9 \\ 0 & 3 & 1 & -1 \end{bmatrix}$$

Row 1 is unchanged

Row 2 is unchanged

Row 1 is added to Row 3

Moving to the second column, we want a 1 in the diagonal position (where there is now –3). We get this by dividing every element in row 2 by -3:

$$\begin{bmatrix} 1 & 1 & -1 & -2 \\ 0 & 1 & -1 & -3 \\ 0 & 3 & 1 & -1 \end{bmatrix}$$

Row 1 is unchanged

Row 2 is divided by –3

Row 3 is unchanged

To obtain a 0 below the 1, we multiply row 2 by -3 and add it to the third row:

$$\begin{bmatrix} 1 & 1 & -1 & -2 \\ 0 & 1 & -1 & -3 \\ 0 & 0 & 4 & 8 \end{bmatrix}$$

Row 1 is unchanged

Row 2 is unchanged

-3 times row 2 is added to row 3

The last step is to obtain a 1 on the diagonal in column 3. We divide each element in Row 3 by 4 to accomplish this:

$$\begin{bmatrix} 1 & 1 & -1 & -2 \\ 0 & 1 & -1 & -3 \\ 0 & 0 & 1 & 2 \end{bmatrix}$$

Row 1 is unchanged

Row 2 is unchanged

Each element in Row 3 is divided by 4

Converting back to a system of equations, we have

x + y – z = -2

y – z = -3

z = 2

We have z = 2. To find y, we back-substitute 2 for z in the second equation:

y – z = -3

y – 2 = -3

y = -1

Finally, we back substitute -1 for y and 2 for z in the first equation:

x + y – z = -2

x + (-1) – 2 = -2

x – 3 = -2

x = 1

The solution is the ordered triple (1, -1, 2).

Example. Solve:

3x – 4y + 4z = 7

x – y – 2z = 2

2x – 3y + 6z = 5

Solution: Start with the augmented matrix:

$$\begin{bmatrix} 3 & -4 & 4 & 7 \\ 1 & -1 & -2 & 2 \\ 2 & -3 & 6 & 5 \end{bmatrix}$$

To obtain a 1 in the top position of the first column, interchange ows 1 and 2:

$$\begin{bmatrix} 1 & -1 & -2 & 2 \\ 3 & -4 & 4 & 7 \\ 2 & -3 & 6 & 5 \end{bmatrix}$$

To get 0's below the 1: multiply row 1 by –3 and add to row 2; multiply row 1 by –2 and add to row 3

$$\begin{bmatrix} 1 & -1 & -2 & 2 \\ 0 & -1 & 10 & 1 \\ 0 & -1 & 10 & 1 \end{bmatrix}$$

To get a 1 in the diagonal position in the second column, multiply row 2 by –1:

$$\begin{bmatrix} 1 & -1 & -2 & 2 \\ 0 & 1 & -10 & -1 \\ 0 & -1 & 10 & 1 \end{bmatrix}$$

Add row 2 to row 3:

$$\begin{bmatrix} 1 & -1 & -2 & 2 \\ 0 & 1 & -10 & -1 \\ 0 & 0 & 0 & 0 \end{bmatrix}$$

Row 3 represents the equation $0 = 0$ which is always true.

This indicates that the system is dependent and has infinitely many solutions.

Row 2 represents the equation $y - 10z = -1$.

Solving this equation for y gives us:

$y = 10z - 1$.

Row 1 represents the equation $x - y - 2z = 2$.

Substitute $10z - 1$ for y and solve for x:

$x - (10z - 1) - 2z = 2$

$x - 10z + 1 - 2z = 2$

$x - 12z + 1 = 2$

$x = 12z + 1$

Therefore, every ordered triple of the form $(12z + 1, 10z - 1, z)$ is a solution of the system.

Moment of Inertia

Moment of inertia, also called mass moment of inertia or the angular mass, (SI units kg m^2) is a measure of an object's resistance to changes in its rotation rate. It is the rotational analog of mass. That is, it is the inertia of a rigid rotating body with respect to its rotation.

The moment of inertia plays much the same role in rotational dynamics as mass does in basic dynamics, determining the relationship between angular momentum and angular velocity, torque and angular acceleration, and several other quantities. While a simple scalar treatment of the moment of inertia suffices for many situations, a more advanced tensor treatment allows the analysis of such complicated systems as spinning tops and gyroscopic motion.

The symbol I and sometimes J are usually used to refer to the moment of inertia.

The concept was introduced by Euler in his book *a Theoria motus corporum solidorum seu rigidorum* in 1730. In this book, he discussed the moment of inertia and many related concepts, such as the principal axis of inertia.

Overview

The moment of inertia of an object about a given axis describes how difficult it is to change its angular motion about that axis. For example, consider two discs, A and B, made of the same material and of equal mass. Disc A is larger in diameter but thinner than B.

It requires more effort to accelerate disc A (change its angular velocity) because its mass is distributed farther from its axis of rotation: mass that is farther out from that axis must, for a given angular velocity, move more quickly than mass closer in. In this case, disc A has a larger moment of inertia than disc B.

The moment of inertia of an object can change if its shape changes. An ice figure skater who begins a spin with arms outstretched provides a striking example. By pulling in her arms, she reduces her moment of inertia, causing her to spin faster (by the conservation of angular momentum).

The moment of inertia has two forms, a scalar form I (used when the axis of rotation is known) and a more general tensor form that does not require knowing the axis of rotation. The scalar moment of inertia I (often called simply the "moment of inertia") allows a succinct analysis of many simple problems in rotational dynamics, such as objects rolling down inclines and the behaviour of pulleys.

For instance, while a block of any shape will slide down a frictionless decline at the same rate, rolling objects may descend at different rates, depending on their moments of inertia. A hoop will descend more slowly than a solid disk of equal mass and radius because more of its mass is located far from the axis of rotation, and thus needs to move faster if the hoop rolls at the same angular velocity.

However, for (more complicated) problems in which the axis of rotation can change, the scalar treatment is inadequate, and the tensor treatment must be used (although shortcuts are possible in special situations).

Examples requiring such a treatment include gyroscopes, tops, and even satellites, all objects whose alignment can change.

The moment of inertia can also be called the mass moment of inertia (especially by mechanical engineers) to avoid confusion with the second moment of area, which is sometimes called the moment of inertia (especially by structural engineers) and denoted by the same symbol I. The easiest way to differentiate these quantities is through their units. In addition, the moment of inertia should not be confused with the polar moment of inertia, which is a measure of an object's ability to resist torsion (twisting).

Scalar Moment of Inertia

Definition

A simple definition of the moment of inertia (with respect to a given axis of rotation) of any object, be it a point mass or a 3D-structure, is given by:

$$I = \int r^2 \, dm$$

where m is mass and r is the perpendicular distance to the axis of rotation.

Detailed Analysis

The (scalar) moment of inertia of a point mass rotating about a known axis is defined by

$$I \stackrel{\Delta}{=} mr^2$$

The moment of inertia is additive. Thus, for a rigid body consisting of N point masses m_i with distances r_i to the rotation axis, the total moment of inertia equals the sum of the point-mass moments of inertia:

$$I \stackrel{\Delta}{=} \sum_{i=1}^{N} m_i r_i^2$$

For a solid body described by a mass density function, ñ(r), the moment of inertia about a known axis can be calculated by integrating the square of the distance (weighted by the mass density) from a point in the body to the rotation axis:

$$I \stackrel{\Delta}{=} \iiint_V \| r \|^2 \rho(r) dV$$

where

V is the volume occupied by the object.

ñ is the spatial density function of the object, and

r = (r, θ, ϕ), (x, y, z), or (r, θ, z) are coordinates of a point inside the body.

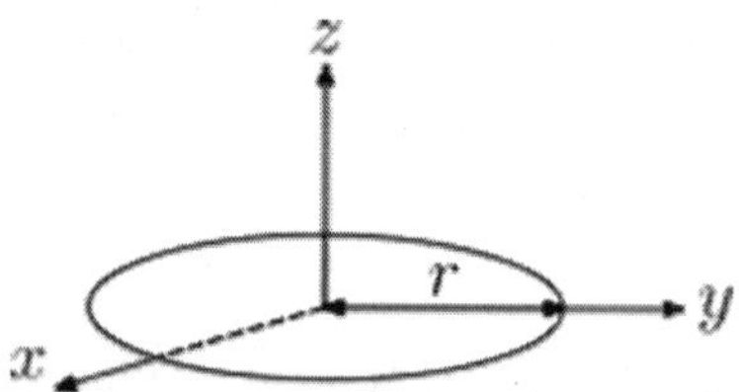

Diagram for the calculation of a disk's moment of inertia. Here k is 1/2 and r is the radius used in determining the moment.

Based on dimensional analysis alone, the moment of inertia of a non-point object must take the form:

$$I = k.M.R^2$$

where

M is the mass

R is the radius of the object from the centre of mass (in some cases, the length of the object is used instead.)

k is a dimensionless constant called the *inertia constant* that varies with the object in consideration.

Inertial constants are used to account for the differences in the placement of the mass from the centre of rotation. Examples include:

- $k = 1$, thin ring or thin-walled cylinder around its centre,
- $k = 2/5$, solid sphere around its centre
- $k = 1/2$, solid cylinder or disk around its centre.

Parallel Axis Theorem

Once the moment of inertia has been calculated for rotations about the centre of mass of a rigid body, one can conveniently recalculate the moment of inertia for all parallel rotation axes as well, without having to resort to the formal definition.

If the axis of rotation is displaced by a distance R from the centre of mass axis of rotation (e.g. spinning a disc about a point on its periphery, rather than through its centre,) the displaced and centre-moment of inertia are related as follows:

$$I_{displaced} = I_{center} + MR^2.$$

This theorem is also known as the *parallel axes rule* and is a special case of *Steiner's parallel-axis theorem.*

Composite Bodies

If a body can be decomposed (either physically or conceptually) into several constituent parts, then the moment of inertia of the body about a given axis is obtained by summing the moments of inertia of each constituent part around the same given axis.

Equations Involving the Moment of Inertia

The rotational kinetic energy of a rigid body can be expressed in terms of its moment of inertia. For a system with N point masses m_i moving with speeds v_i, the rotational kinetic energy T equals

$$T = \sum_{i=1}^{N} \frac{1}{2} m_i v_i^2 = \sum_{i=1}^{N} \frac{1}{2} m_i (\omega r_i)^2 = \frac{1}{2} \sum_{i=1}^{N} m_i r_i^2 \omega^2 = \frac{1}{2} I \omega^2$$

where ù is the common angular velocity (in radians per second). The final expression $I\omega^2 / 2$ also holds for a mass density function with a generalization of the above derivation from a discrete summation to an integration.

In the special case where the angular momentum vector is parallel to the angular velocity vector, one can relate them by the equation

$$L = I\omega$$

where L is the angular momentum and ù is the angular velocity. However, this equation does not hold in many cases of interest, such as the torque-free precession of a rotating object, although its more general tensor form is always correct.

When the moment of inertia is constant, one can also relate the torque on an object and its angular acceleration in a similar equation:

$$\tau = I\alpha$$

where τ is the torque and α is the angular acceleration.

Derivation of the Tensor Components

The distance r of a particle at x from the axis of rotation passing through the origin in the $\hat{n}$ direction is $|x-(x.\hat{n})\hat{n}|$.

By using the formula $I = mr^2$ (and some simple vector algebra) it can be seen that the moment of inertia of this particle (about the axis of rotation passing through the origin in the direction) is This is a quadratic form in and, after a bit more algebra, this leads to a tensor formula for the moment of inertia

$$I = m[n_1, n_2, n_3]\begin{bmatrix} y^2+z^2 & -xy & -xz \\ -yx & x^2+z^2 & -yz \\ -zx & -zy & x^2+y^2 \end{bmatrix}\begin{bmatrix} n_1 \\ n_2 \\ n_3 \end{bmatrix}.$$

This is exactly the formula given below for the moment of inertia in the case of a single particle. For multiple particles we need only recall that the moment of inertia is additive in order to see that this formula is correct.

Reduction to Scalar

For any axis, represented as a column vector with elements n_i, the scalar form I can be calculated from the tensor form I as

$$I = \hat{n}^T I \hat{n} = \sum_{j=1}^{3}\sum_{k=1}^{3} n_j I_{jk} n_k.$$

The range of both summations correspond to the three Cartesian coordinates.

The following equivalent expression avoids the use of transposed vectors which are not supported in maths libraries because internally vectors and their transpose are stored as the same linear array,

$$I = I^T \hat{n}.\hat{n}.$$

However it should be noted that although this equation is mathematically equivalent to the equation above for any matrix, inertia tensors are symmetrical. This means that it can be further simplified to:

$$I = I\hat{n}.\hat{n}.$$

Principal Moments of Inertia

By the spectral theorem, since the moment of inertia tensor is real and symmetric, it is possible to find a Cartesian coordinate system in which it is diagonal, having the form

$$\mathrm{I} = \begin{bmatrix} I_1 & 0 & 0 \\ 0 & I_2 & 0 \\ 0 & 0 & I_3 \end{bmatrix}$$

where the coordinate axes are called the principal axes and the constants I_1, I_2 and I_3 are called the principal moments of inertia. The unit vectors along the principal axes are usually denoted as (e_1, e_2, e_3). This result was first shown by *J. J. Sylvester (1852)*, and is a form of Sylvester's law of inertia.

When all principal moments of inertia are distinct, the principal axes are uniquely specified. If two principal moments are the same, the rigid body is called a symmetrical top and there is no unique choice for the two corresponding principal axes. If all three principal moments are the same, the rigid body is called a spherical top (although it need not be spherical) and any axis can be considered a principal axis, meaning that the moment of inertia is the same about any axis.

The principal axes are often aligned with the object's symmetry axes. If a rigid body has an axis of symmetry of order *m*, i.e., is symmetrical under rotations of 360°/*m* about a given axis, the symmetry axis is a principal axis. When $m > 2$, the rigid body is a symmetrical top. If a rigid body has at least two symmetry axes that are not parallel or perpendicular to each other, it is a spherical top, e.g., a cube or any other Platonic solid.

A practical example of this mathematical phenomenon is the routine automotive task of balancing a tire, which basically means adjusting the distribution of mass of a car wheel such that its principal axis of inertia is aligned with the axle so the wheel does not wobble.

Parallel Axis Theorem

Once the moment of inertia tensor has been calculated for rotations about the centre of mass of the rigid body, there is a useful labour-saving method to compute the tensor for rotations offset from the centre of mass.

If the axis of rotation is displaced by a vector R from the centre of mass, the new moment of inertia tensor equals

$$I^{displaced} = I^{center} + M[(R.R)E_3 - R \otimes R]$$

where M is the total mass of the rigid body, E_3 is the 3 × 3 identity matrix, and is the outer product.

Rotational Symmetry

Using the above equation to express all moments of inertia in terms of integrals of variables either along or perpendicular to the axis of symmetry usually simplifies the calculation of these moments considerably.

Other Mechanical Quantities

Comparison With Covariance Matrix

The moment of inertia tensor about the centre of mass of a 3 dimensional rigid body is related to the covariance matrix of a trivariate random vector whose probability density function is proportional to the pointwise density of the rigid body by:

$$I = n(1_{3\times3} tr(\Sigma) - \Sigma)$$

where n is the number of points.

The structure of the moment-of-inertia tensor comes from the fact that it is to be used as a bilinear form on rotation vectors in the form

$$\frac{1}{2}\omega^T I \omega.$$

Each element of mass has a kinetic energy of

$$\frac{1}{2}m|v|^2.$$

The velocity of each element of mass is $\omega \times r$ where r is a vector from the centre of rotation to that element of mass. The cross product can be converted to matrix multiplication so that

$$\omega \times r = [r]^T \times \omega$$

and similarly

$$(\omega \times r)^T = ([r] \times \omega)^T = \omega^T [r]_\times^T.$$

Thus

$$|v|^2 = (\omega \times r)^T (\omega \times r) = \omega^T [T]_\times^T [r] \times \omega$$

plugging in the definition of $[.]_\times$ the $[r]_\times^T [r]_\times$ term leads directly to the structure of the moment tensor.

Second Moment of Area

The second moment of area, also known as the area moment of inertia or second moment of inertia is a property of a shape that can be used to predict the resistance of beams to bending and deflection.

The deflection of a beam under load depends not only on the load, but also on the geometry of the beam's cross-section.

This is why beams with higher area moments of inertia, such as I-beams, are so often seen in building construction as opposed to other beams with the same area.

It is analogous to the polar moment of inertia, which characterizes an object's ability to resist torsion.

The *second moment of area* is not the same thing as the moment of inertia, which is used to calculate angular acceleration. Many engineers refer to the *second moment of area* as the *moment of inertia* and use the same symbol I for both, which may be confusing.

Which inertia is meant (accelerational or bending) is usually clear from the context and obvious from the units: second moment of area has units of length to the fourth power whereas moment of inertia has units of mass times length squared.

Notes on Notation

In some circumstances I_x and I_{xx} mean the same thing depending on what notation the person conducting the calculations is using.

For example some people referring to a section across the x-axis will just use I_x, but someone who refers to a section across the x-axis as 'section x-x' will use I_{xx} for the same variable.

The same applies in the case of I_y and I_{yy}.

Definition

$$I_x = \int y^2 dA$$

- I_x = the second moment of area about the axis x
- dA = an elemental area
- y = the perpendicular distance from the axis x to the element dA

Intuitively, the second moment of area, measuring the resistance to bending, can be likened to a person's attempt to stop a force from turning a lever: the farther a person places a hand from the pivot the more leverage is obtained and the easier it is to resist the turning force.

In the above formula, the hands resisting turning are replaced with the sum of small sections of the object (infinitesimally small in the limit); the leverage is proportional to the square of the distance from the 'pivot'.

Each small section adds its own contribution depending on its position and proportional to how big it is in cross section; each piece can be split into smaller pieces being summed up until the infinitesimal size is reached and the result is accurate (i.e. the limit of the integral).

The above can only be used on its own when sections are symmetrical about the x-axis. When this is not the case, the second moment of area about both the x- and the y-axis and the product moment of area, I_{xy}, are required.

Product Moment of Area

The product moment of area, I_{xy} is defined as

$$I_{yx} = \int xy dA$$

- dA = an elemental area
- x = the perpendicular distance to the element dA from the axis y
- y = the perpendicular distance to the element dA from the axis x

Note that I_{xy} is defined here with positive sign following common practice adopted in structural analysis.

The alternative approach to definition of moments of area could be to define the symmetric tensor for moment of area similarly to the tensor of mass moment of inertia; note that non-diagonal members of this tensor are equal to $-I_{xy}$:

$$J_{ij} = \int (r^2 \delta Lij - r_i r_j) dA$$

The product moment of area is significant for determining the bending stress in an asymmetric cross section. Unlike the second moments of area, the product moment may give both negative and positive values.

A coordinate system, in which the product moment is zero, is referred to as a set of principal axes, and the second moments of area calculated with respect to the principal axes will assume their maxima and minima. A coordinate system with origin in the centroid of the cross section and with both axes being axes of symmetry are always principal axes.

Unit

The unit for second moment of area is length to the fourth power (typically mm^4, in^4, etc.)

Co-ordinate Transformations

When calculating moments of the section it is often practical to compute them in one co-ordinate system (typically bound to the section shape) and then transform to another one using co-ordinate transformations.

Parallel Axis Theorem

The parallel axis theorem can be used to determine the moment of an object about any axis, given the moment of inertia (second moment of area) of the object about the parallel axis through the object's centre of mass (or centroid) and the perpendicular distance between the axes.

$$I_x = I_{xCG} + Ad^2$$

- I_x = the second moment of area with respect to the x-axis
- I_{xCG} = the second moment of area with respect to an axis parallel to x and passing through the centroid of the shape (coincides with the neutral axis)
- A = area of the shape
- d = the distance between the x-axis and the centroidal axis

Axis Rotation

The following formulae can used to calculate moments of the section in a co-ordinate system rotated relative to the original coordinate system:

$$I_x * = \frac{I_x + I_y}{2} + \frac{I_x - I_y}{2}\cos(2\phi) - I_{xy}\sin(2\phi)$$

$$I_x * = \frac{I_x + I_y}{2} - \frac{I_x - I_y}{2}\cos(2\phi) + I_{xy}\sin(2\phi)$$

$$I_{xy} * = \frac{I_x - I_y}{2}\sin(2\phi) + I_{xy}\cos(2\phi)$$

- ϕ = the angle of rotation (anticlockwise sense):

  $x^* = x\cos\phi + y\sin\phi$

  $y^* = -x\sin\phi + y\cos\phi$
- I_x, I_y and I_{xy} = the second moments and the product moment of area in the original coordinate system
- I_x^*, I_y^* and I_{xy}^* = the second moments and the product moment of area in the rotated coordinate system.

The value of the angle ϕ, which will give a product moment of zero, is equal to:

$$\phi = -\frac{1}{2}\arctan\frac{2I_{xy}}{I_x - I_y}$$

This angle is the angle between the axes of the original coordinate system and the principal axes of the cross section.

Stress in a Beam

The general form of the classic bending formula for a beam in coordinate system having origin located at the neutral axis of the beam is (Pilkey 2002, p. 17):

$$\sigma = -\frac{M_y I_x + M_x I_{xy}}{I_x I_y - I_{xy}^2}x + \frac{M_x I_y + M_y I_{xy}}{I_x I_y - I_{xy}^2}y$$

- σ is the normal stress in the beam due to bending
- x = the perpendicular distance to the centroidal y-axis
- y = the perpendicular distance to the centroidal x-axis
- M_y = the bending moment about the y-axis

- M_x = the bending moment about the x-axis
- I_x = the second moment of area about x-axis
- I_y = the second moment of area about y-axis
- I_{xy} = the product moment of area.

If the coordinate system is chosen to give a product moment of area equal to zero, the formula simplifies to:

$$\sigma = -\frac{M_y}{I_y}x + \frac{M_x}{I_x}y$$

If additionally the beam is only subjected to bending about one axis, the formula simplifies further:

$$\sigma = \frac{M}{I_x}y$$

Second Moment of Area for Various Cross Sections

Rectangular Cross Section

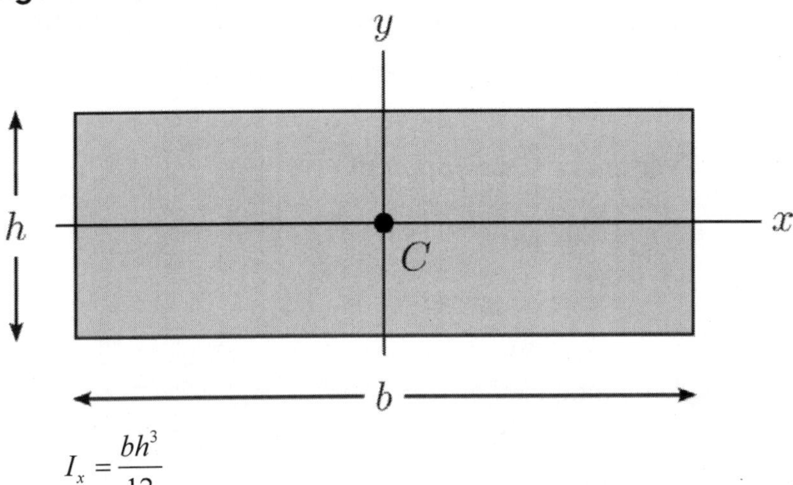

$$I_x = \frac{bh^3}{12}$$

- b = width (x-dimension),
- h = height (y-dimension)

$$I_y = \frac{hb^3}{12}$$

- b = width (x-dimension),
- h = height (y-dimension)

Circular Cross Section

$$I_0 = \frac{\pi}{64}D^4 = \frac{\pi}{4}r^4$$

- D = diameter
- r = radius

this equation is very useful in mast calculations.

Having given moment of inertia calculated previously and entering it to this equation, gives you rapid approximation of required mast diameter.

Whats more regarding most calculations, you can easily check whether your moment of inertia of mast will be sufficient for a given maximum load, by calculating from the Euler's formula:

$$F = \frac{\pi^2 EI}{l_2}$$

- E = Young (elastic) modulus of material [N/mm2] (ex. for aluminium 70500 N/mm^2 carbon fibre UNI 120000 N/mm^2)
- I = second moment of area of examined object [mm^4]
- l = length of panel [mm].

Hollow Cylindrical Cross Section

$$I_0 = \frac{\pi}{64}(D_O^4 - D_I^4) = \frac{\pi}{4}(r_O^4 - r_I^4)$$

- D_O = outside diameter
- D_I = inside diameter
- r_O = outside radius
- r_I = inside radius.

Composite Cross Section

When it is easier to compute the moment for an item as a combination of pieces, the second moment of area is calculated by applying the parallel axis theorem to each piece and adding the terms:

$$I_x = \sum(I_{local} + y^2 A)$$

$$I_y = \sum(I_{local} + x^2 A)$$

- y = distance from x-axis
- x = distance from y-axis
- A = surface area of part
- I_{local} is the second moment of area for that part of the composite, in the appropriate direction (i.e. I_x or I_y respectively).

"I-beam" Cross Section

I-beam

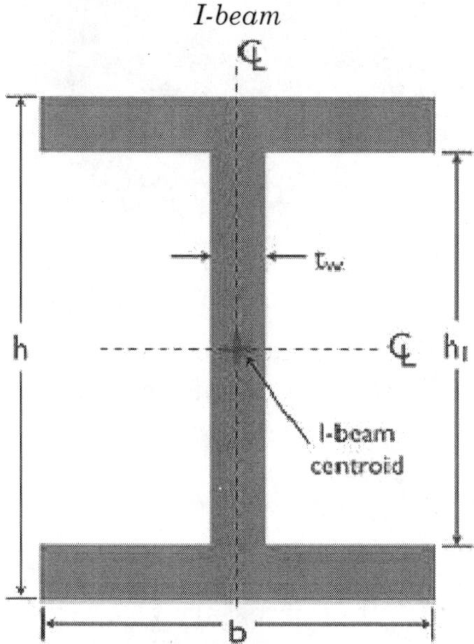

The I-beam can be analysed as either three pieces added together or as a large piece with two pieces removed from it.

Either of these methods will require use of the formula for composite cross section.

This section only covers *doubly symmetric* I-beams, meaning the shape has two planes of symmetry.

- b = width (x-dimension),
- h = height (y-dimension)
- t_w = width of central webbing
- h_1 = inside distance between flanges (usually referred to as h_w, the height of the web)

This formula uses the method of a block with two pieces removed. (While this may not be the easiest way to do this calculation, it is instructive in demonstrating how to subtract moments).

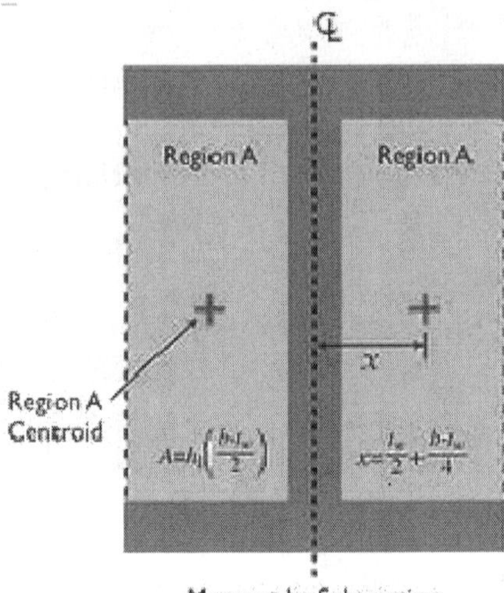

Figure: *I-beam Diagram, Moment by Subtraction*

Since the I-beam is symmetrical with respect to the y-axis the I_x has no component for the centroid of the blocks removed being offset above or below the x axis.

$$I_x = \frac{bh^3 - 2\frac{b-t_\omega}{2}h_1^3}{12}$$

When computing I_y it is necessary to allow for the fact that the pieces being removed are offset from the Y axis, this results in the Ax^2 term.

$$I_y = \frac{hb^3}{12} - 2\left(\frac{h_1\left(\frac{b-t_\omega}{2}\right)^3}{12} + Ax^2\right)$$

- A = Area contained within the middle of one of the 'C' shapes of created by two flanges and the webbing on one side of the cross section =
- x = distance of the centroid of the area contained in the 'C' shape from the y-axis of the beam $= h_1 \frac{b-t_w}{2}$

Doing the same calculation by combining three pieces, the centre webbing plus identical contributions for the top and bottom piece:

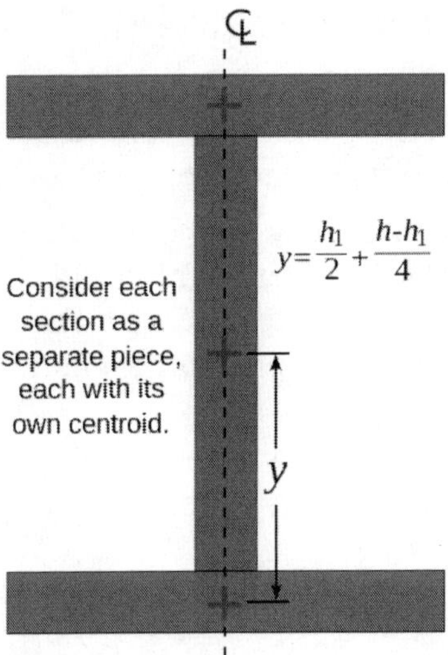

***Figure:** I-beam Diagram, Moment by Addition*

Since the centroids of all three pieces are on the y-axis I_y can be computed just by adding the moments together.

$$I_y = \frac{h_1 t_w^{\ 3}}{12} + 2\frac{\frac{h-h_1}{2}b^3}{12}$$

However, this time the law for composition with offsets must be used for I_x because the centroids of the top and bottom are offset from the centroid of the whole I-beam.

- A = Area of the top or bottom piece $=b\frac{h-h_1}{2}$
- y = offset of the centroid of the top or bottom piece from the centroid of the whole I-beam $=\frac{h+h_1}{4}$

$$I_x = \frac{t_w h_1^{\ 3}}{12} + 2\left(\frac{b(h-h_1)^3}{12} + Ay^2\right)$$

$$= \frac{t_w h_1^{\ 3}}{12} + 2\left(\frac{b\left(\frac{h-h_1}{2}\right)^3}{12} + b\frac{h-h_1}{2}\left(\frac{h+h_1}{4}\right)^2\right)$$

Any Cross Section Defined as Polygon

The second moments of area for any cross section defined as a simple polygon on XY plane can be computed in a generic way by summing contributions from each segment of a polygon.

For each segment defined by two consecutive points of the polygon, consider a triangle with two corners at these points and third corner at the origin of the coordinates. Integration by the area of that triangle and summing by the polygon segments yields:

$$I_x = \frac{1}{12}\sum_{i=1}^{n}(y_i^2 + y_i y_{i+1} + y_{i+1}^2)a_i$$

$$I_y = \frac{1}{12}\sum_{i=0}^{n}(x_i^2 + x_i x_{i+1} + x_{i+1}^2)a_i$$

$$I_{xy} = \frac{1}{12}\sum_{i=1}^{n}(x_i y_{i+1} + 2x_i y + 2x_i +_{1yi+1} + x_{i+1yi})a_i$$

- $a_i = x_i y_{i+1} - x_{i+1} y_i$ is twice the (signed) area of the elementary triangle,
- index i passes over all n points in the polygon, which is considered closed, i.e. point $n+1$ is point *1*

These formulae imply that points defining the polygon are ordered in anticlockwise manner; for clockwisely defined polygons it will give negative values.

5

Applications of Trigonometry

The Law of Sines

Given three non-collinear points, A, B and C, there is only one circle containing those three points. Three non-collinear points also form the vertices of a triangle. According to a theorem from geometry, the ratio of a side of a triangle to the sine of the opposite angle equals the diameter of the circumscribed circle. Since for a given triangle there is only one circumscribed circle, then one may compute its diameter by computing the ratio between *any* of its three sides and the sine of the angle opposite that side. That is,

$$d = a \,/\, \sin \text{A} = b \,/\, \sin \text{B} = c \,/\, \sin \text{C}.$$

This identity is called the Law of Sines.

The Law of Sines, together with the Law of Cosines which we will study shortly, are useful for solving triangles.

There are six pieces of information that must be known before a triangle is 'solved': the size of its three angles and the size of its three sides. In order to solve a triangle, one needs to have at least three of the six pieces of information to begin with. The one exception to this principle is this: knowing only the three angles is insufficient for finding the three sides

Why can't one determine the length of the three sides of a triangle if one knows all three angles?

Solution

There is such a thing as *similar triangles*: triangles of different sizes, yet which have congruent angles. For example, two equilateral triangles of different sizes still have all their angles equal to 60°. So just knowing the angles tells us nothing about the size of the sides.

There are six different ways to classify the three given pieces of information:

SSS: The lengths of all three sides are given.

SAS: The lengths of two sides and the size of the angle between them are given.

SSA: The lengths of two sides and the size of the angle opposite one of them are given.

ASA: The sizes of two angles and the length of the side between them are given.

SAA: The sizes of two angles and the length of the side opposite one of them are given.

AAA: The sizes of all three angles are given.

Note that some combinations of three pieces of information will not have a solution. For example, the sum of the three angles of a triangle must be 180°.

Thus in the case of ASA, if the two angles add to more than 180°, there will be no solution. Or in the case of SSS, if the sum of two of the sides is less than the third side, there will be no solution, since the sum of two sides of a triangle is always greater than the third side. SAS is the only case where all possible lengths of the sides and all possible angles greater than 0° and less than 180° will have a solution.

Exercise : Explain why each of the following 'triangles' has no solution:

(a) SSS: $a = 7,\ b = 2,\ c = 4$

(b) ASA: $A = 85°,\ b = 10,\ C = 98°$

(c) SAA: $a = 12,\ B = 100°,\ C = 90°$

(d) AAA: $A = 65°,\ B = 82°,\ C = 35°$

Solution

(a) The sum of two sides of a triangle must be larger than the third. Bud $b + c = 6$ and $a = 7$. So there is no solution.

(b) The sum of two angles of a triangle must be smaller than 180°. But $A + B = 183°$, so there is no solution.

(c) The sum of B and C is greater than 180° so there is no solution.

(d) The sum of the three angles is greater than 180°, so there is no solution.

In order to apply the Law of Sines, it is necessary to know the size of at least one side and the size of the angle opposite that side. That is, one must know at least one side-sine pair.

One will know a side-sine pair in the case of SSA or SAA. But if one is given ASA, one can deduce the third angle since the sum all three angles must be 180°. Thus one will know a side-sine pair given ASA.

So the Law of Sines is used to solve triangles given SSA, SAA or ASA.

SAA and ASA are worked essentially the same: find the third angle, then use the side-sine pair to solve the other two sides.

Example: Given a = 10, A = 30° and C = 120°, solve the triangle.

First find B = 180° – 30° – 120° = 30°. Next, use the fact that b/sin B = a/sin A to find b, since $b = a$ sin B/sin A = 10 sin (30°)/sin (30°) = 10.

Likewise, c/sin C = a/sin A, so $c = a$ sin C/sin A = 10 sin (120°)/sin (30°)

$$= 10\left(\sqrt{3}/2\right)(1/2) = 5\sqrt{3}/2.$$

Thus the solution of the triangle is:

B = 120°

b = 10

$c = 5\sqrt{3/2}$

Exercise

Given A = 24°, B = 80° and c = 20, solve the triangle.

Solution

A = 24°, B = 80°, c = 20

At first glance it might appear that we do not have a side-sine pair. But we can easily get the angle C opposite the given side and then apply the Law of Sines. This problem is of the type SAA.

C = 180° - 24° - 80° = 76°

We use the law of sines to find sides a and b.

$$\frac{a}{\sin 24^0} = \frac{20}{\sin 76^0}$$

$$a = \frac{20 \sin 24^0}{\sin 76^0} = 8.38$$

$$\frac{b}{\sin 80^0} = \frac{20}{\sin 76^0}$$

$$b = \frac{20 \sin 80^0}{\sin 76^0} = 20.30$$

It is possible for the case SSA to have no solution for some combinations of lengths of the sides and angles.

Exercise

Explain why the following SSA 'triangle' has no solution. Hint: the sine of an angle cannot be greater than 1.

$$A = 60^o,\ a = 2,\ b = 10$$

Solution

$$A = 60^\circ,\ a = 2,\ b = 10$$

$$\frac{\sin B}{10} = \frac{\sin 60}{2} = \frac{\sqrt{3}}{4}$$

$$\sin B = \frac{5\sqrt{3}}{2} \approx 4.33 > 1$$

Since the sine of an angle cannot exceed 1, there can be no solution to this triangle.

Knowing the sine of one of the angles of a triangle is not sufficient information to uniquely determine the angle.

For example, suppose we want to find A and all we know is that sin A = 1/2. Then A could be 30° or A could be 150°. We cannot assume that the correct value is 30° unless we can rule out 150°.

In general, if sin A = y, then either A = arcsin y or A = 180° - arcsin y.

Sometimes this situation arises when solving SSA triangles. When this happens and when we cannot rule out the larger solution, then we must list *two separate solutions* to the triangle.

Exercise

Find *two* solutions to the triangle:

$a = 10,\ b = 4,\ B = 20^\circ$

Solution

$a = 10,\ b = 4,\ B = 20^\circ$

$$\frac{\sin A}{10} = \frac{\sin 20^\circ}{4}$$

$$\sin A = \frac{10 \sin 20^\circ}{4} \approx 0.8550$$

There are two angles between 0° and 180° whose sine is 0.8550:

58.8° and 121.2°.

Since the larger of the two angles, when added to the given angle 20° is only 141.2°, which is still less than 180°, the larger angle is a possible solution. There will be two different solutions to this triangle.

Solution #1

$A = 58.8^\circ$

$C = 180^\circ - 20^\circ - 58.8^\circ = 101.2^\circ$

$$\frac{c}{\sin 101.2^0} = \frac{4}{\sin 20^0}$$

$$c = \frac{4 \sin 101.2^0}{\sin 20^0}$$

Solution #2

$A = 121.2^\circ$

$C = 180^\circ - 20^\circ - 121.2^\circ = 38.8^\circ$

$$\frac{c}{\sin 38.8^0} = \frac{4}{\sin 20^0}$$

$$c = \frac{4 \sin 38.8^0}{\sin 20^0} = 7.3$$

Exercise

A ship sailing due north spots a lighthouse 30° to the left of its line of travel. Two miles later, the lighthouse is 50° to the left of its line of travel.

How far is the ship from the lighthouse at that point? (Assume the earth is flat.)

Solution

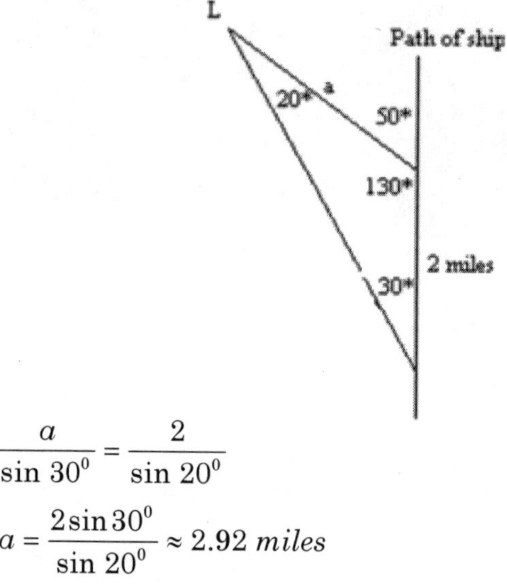

$$\frac{a}{\sin 30^0} = \frac{2}{\sin 20^0}$$

$$a = \frac{2\sin 30^0}{\sin 20^0} \approx 2.92\ miles$$

Exercise

Scientists wish to measure the diameter of a large circular meteor crater. Two points A and B on the edge of the pit are 120 feet apart. From a point C on the far side of the pit, the angle between the lines AC and BC is measured to be 8°. Use the Law of Sines to find the diameter of the crater.

Solution

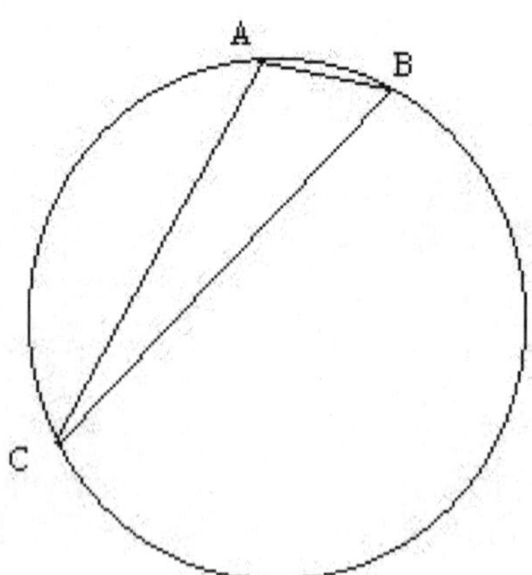

Remember, the Law of Sines states that the diameter of the circumscribed circle for a triangle ABC equals any side of the triangle divided by the sine of the opposite angle.

Since we know the side AB has length 120 feet and the angle opposite the side AB is 8°, then the diameter must be

$$D=\frac{120}{\sin 8^0}=862.2 \text{ feet}$$

The Law of Cosines

The Law of Cosines is a generalization of the Pythagorean Theorem. In a triangle, if C is a right angle, then $a^2 + b^2 = c^2$. The Law of cosines states that

$$a^2 + b^2 - 2\ ab \cos \text{C} = c^2.$$

But if C is not 90°, then that angle has lost its special Status. Thus the Law of Cosines must apply to other given variables.

So

$$b^2 + c^2 - 2\ bc \cos \text{A} = a^2 \text{ and}$$

$$a^2 + c^2 - 2\ ac \cos \text{B} = b^2.$$

The Law of Cosines may be used to solve triangles of type SAS and SSS.

To solve SAS problems, use the Law of Cosines to find the length of the side opposite the given angle.

Example:, given $a = 10$, B = 32° and $c = 15$.

The first step is to find the side opposite the given angle, in this case, b.

$$b^2 = a^2 + c^2 -2\ ac \cos \text{B. Thus,}$$

$$b^2 = 100 + 225 - 2\ (10)\ (15)\cos 32^\circ$$

$$= 325 - 300\ (0.848048096) = 325 - 254.4 = 70.59.$$

Thus $b = 8.40$, approximately.

Now, we must find the sizes of angles A and C. We can use the Law of Sines, *provided we find the angle opposite the smaller side first.*

Why is it important to find the angle opposite the smaller of a and c first? If the triangle has an obtuse angle (i.e. > 90°), then it will be opposite the largest side of the triangle.

So we know that the angle opposite the smaller of the two given sides *cannot* be obtuse. Why should we be concerned whether an angle is obtuse or not? Remember that knowing the sine of an angle of a triangle does not tell you what the angle is, since there are *two solutions* to an equation sin D = y between 0° and 180°, an acute angle D = arcsin y and an obtuse angle D = 180° - arcsin y.

If we know in advance that the angle D is acute, then we know to use D = arcsin y.

In the example, we know a = 10, B = 32°, c = 15 and b = 8.40. Since a is smaller than c, we find A next using the Law of Sines.

sin A/10 = sin 32°/8.40

sin A = 0.6309

A = arcsin 0.6309 = 39.1°

Then the remaining angle can be found

C = 180° - 32° - 39.1° = 108.9°

So the solution is b = 8.40, A = 39.1°, C = 108.9°.

Exercise

Given A = 28°, b = 14, c = 10 solve the triangle.

Solution

A = 28°, b = 14, c = 10

Since we do not have a side-sine pair, we must use the Law of Cosines. This problem is of the SAS type: we are given two sides and the included angle (the angle between the two given sides).

First, we will solve for the side opposite the given angle :

$$a^2 = b^2 + c^2 - 2bc\cos A = 196 + 100 - 280\cos(28^0) \approx 48.77$$

$$a = 6.98$$

Next we will use the Law of Sines to solve the angle opposite the *smaller* of the two given sides. It's important that we solve for the angle opposite the smaller of the two sides, since we know that it will not be obtuse.

The angle opposite the largest side of a triangle may be an obtuse angle, that is, it may be greater than 90°.

It is best to find obtuse angles last by subtracting the other two angles from 180°.

$$\frac{\sin C}{10} = \frac{\sin 28^0}{6.98}$$

$$\sin C = \frac{10 \sin 28^0}{6.98} - 0.6722$$

$$C \approx 42.2^0$$

$$B = 180° - 28° - 42.2° = 109.8°$$

Exercise

An engineer wishes to measure the diameter of a hole in the ground, but his tape measure isn't long enough. He places stakes at points A and B on opposite sides of the hole, then places a third stake at a point C at the edge of the hole somewhere between A and C.

His tape is long enough to measure the distance from C to A at 28 feet and the distance from C to B at 26 feet. From the point C, the angle between the lines AC and CB is 158°. How far is it from A to B?

Solution

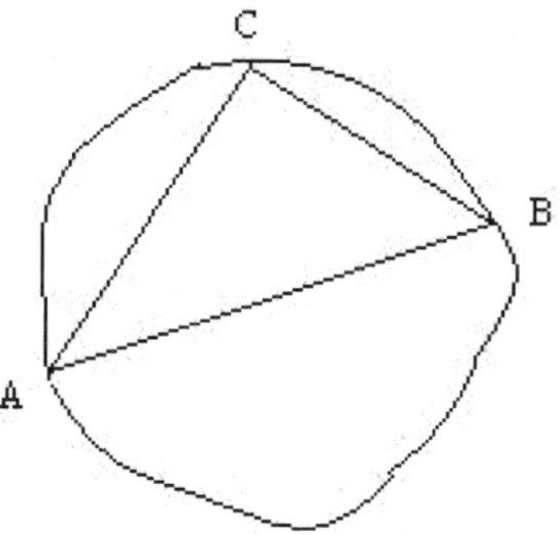

$$|AB|^2 = 28^2 + 26^2 - 2\,(28)(26)\cos 158° = 2810.0$$

So, | AB | H" 53 feet

There is an alternate form of the Law of Cosines which is used to solve the SSS case. In the standard form of the Law of Cosines, each of the three equations is solved for the cosine of the angle to get

$\cos A = (b^2 + c^2 - a^2)/(2\,bc)$

$\cos B = (a^2 + c^2 - b^2)/(2\,ac)$

$\cos C = (a^2 + b^2 - c^2)/(2\,ab)$

Thus, given the lengths of the three *sides* of a triangle, we can find the cosines of the three *angles*.

Knowing the cosine of an angle of a triangle, can we deduce the angle itself? If we know that cos D = x, where D is an interior angle of a triangle, can we deduce that D = arccos x ?

Recall that the principle domain of the cosine function (for the purpose of finding the inverse function) is the interval [0, 180°]. This means that there is *only one* angle between 0° and 180° whose cosine is x. Thus, we can conclude that

$$A = \arccos[\ (b^2 + c^2 - a^2)/(2\ bc)\ ]$$

$$B = \arccos[\ (a^2 + c^2 - b^2)/(2\ ac)\ ]$$

$$C = \arccos[\ (a^2 + b^2 - c^2)/(2\ ab)\ ]$$

Typically, we use the Law of Cosines to find A and B, then subtract from 180° to find C.

Exercise

Given $a = 8$, $b = 5$ and $c = 4$, solve the triangle.

Solution

$$A = \arccos\left(\frac{b^2 + c^2 - a^2}{2bc}\right) = \arccos\left(-\frac{23}{40}\right) = 125.1^0$$

$$B = \arccos\left(\frac{a^2 + c^2 - b^2}{2ac}\right) = \arccos\left(-\frac{55}{64}\right) = 30.8^0$$

$$C = 180^0 - 125.1^0 - 30.8^0 = 24.1^0$$

There is a formula from antiquity for computing the area of a triangle from the lengths of its three sides called Heron's Formula (also called Hero's Formula).

I makes use of the idea of the *semiperimeter*. The semiperimeter is simply *half* the perimeter.

Given a triangle ABC with sides a, b and c and semiperimeter s, the area can be given as: Area = $[\ s\ (s - a)(s - b)(s - c)\ ]^{1/2}$

Exercise

Given a triangle ABC with sides $a = 8$, $b = 5$ and $c = 4$, find the area of the triangle.

Solution

$$s = (1/2)\ (a + b + c) = 8.5$$

$$\text{Area} = -\ [\ s\ (s - a)(s - b)(s - c)\ ] = 9.045$$

Trigonometry and Complex Numbers

Let $z = a + b\, i$ denote a complex number with real part a and imaginary part b. Then the absolute value of z, denoted $|z|$, is the distance from z to zero in the complex plane and is computed by the formula $|z| = \sqrt{a^2 + b^2}$

The line containing 0 and z makes an angle t with the real axis satisfying

$\tan(t) = b/a$. Furthermore,

$\cos(t) = a/|z|$ and

$\sin(t) = b/|z|$. Thus

$a = |z| \cos(t)$ and $b = |z| \sin(t)$, so

$z = |z| \cos(t) + |z| \sin(t)\, i$. Thus

$z = |z|\, [\cos(t) + i \sin(t)]$

This is called the *trigonometric form* of a complex number.

Suppose z1 and z2 are complex numbers. Then their product is

$$z_1 \cdot z_2 = |z_1| [\cos(t_1) + i \sin(t_1)] |z_2| [\cos(t_2) + i \sin(t_2)]$$

$$= |z_1| |z_2| \{[\cos(t_1)\cos(t_2) - \sin(t_1)\sin(t_2)]$$

$$+ i [\cos(t_1) \sin(t_2) + \sin(t_1) \cos(t_2)]\}$$

$$= |z_1| |z_2| [\cos(t_1 + t_2) + i \sin(t_1 + t_2)]. \text{ Thus,}$$

$$z_1 \cdot z_2 = |z_1| |z_2| [\cos(t_1 + t_2) + i \sin(t_1 + t_2)]$$

Exercise

Let $z = 3 + 3\, i$. Find $|z|$. Find an angle t such that $z = |z|$ $[\cos(t) + i \sin(t)]$. Write z in trigonometric form.

Solution

The magnitude of z is found by taking the square root of the sum of the squares if its components, thus:

$|z| = 3\sqrt{2}$

The angle is the standard angle with terminal side through the point (3, 3), or $\pi/4$. Thus, the trigonometric form for the complex number $3 + 3\, i$ is:

$3\sqrt{2}\, [\cos(\pi/4) + i \sin(\pi/4)]$

Exercise

Show that $z^2 = | z |^2 [\cos (2 t) + i \sin (2 t)]$.

Solution

$z = | z | (\cos t + i \sin t)$

So $z^2 = |z|^2 [\cos (t + t) + i \sin (t + t)] = | z |^2 [\cos (2 t) + i \sin (2 t)]$

The result in Exercise 6.3.2 can be extended to the general result, called DeMoivre's Theorem:

If $z = | z | [\cos (t) + i \sin (t)]$, then $z^n = | z |^n [\cos (n t) + i \sin (n t)]$.

Example: Find $(-1 + 2 i)^5$

Let $z = -1 + 2 i$. Then find z^5. First, we must find $|z| = \sqrt{1^2 + 2^2} = \sqrt{5}$ which is approximately 2.236067977. Next, we must find the angle t.

Since z lies in the second quadrant of the complex plane, we are looking for an angle t in the second quadrant satisfying $\tan (t) = b/a = -2$. arctan (–2) will not do, since it is in quadrant IV.

We want pi + arctan (–2), which is approximately 2.034443936. [Your calculator must be in radian mode.] Thus, $(-1 + 2 i)^5$ is approximately equal to

$$(2.236067977)^5 [\cos 10.17221968 + \mathrm{i} \sin 10.17221968]$$
$$= 55.90169937 [-0.733430296 - 0.679764665 i]$$
$$= - 40.99999992 - 37.99999999 i.$$

Since the real and imaginary parts of $(-1 + 2 i)^5$ must be integers, the answer is $(-1 + 2 i)5 = - 41 - 38 i$.

Exercise

Find $(1 + i)^7$.

Solution

First we must put $(1 + i)$ in trigonometric form. The magnitude is $\sqrt{2}$ and the angle is 45°. The seventh power of the complex number will have a magnitude which is the seventh power of $\sqrt{2}$, or $8\sqrt{2}$. The angle of the seventh power will be seven times the angle 45°, or 315°. Thus, the seventh power of $(1 + i)$ is

$$8\sqrt{2} (\cos 315° + i \sin 315°) = 8 - 8 i$$

The inverse of finding powers of complex numbers is finding roots of complex numbers. A complex number has two square roots, three cube roots, four fourth roots, etc. Generally, a complex number has n nth roots. All n of the nth roots of a complex number z are evenly spaced around a circle centred at 0 and having a radius equal to the nth root of the absolute value of z.

Furthermore, if $z = |z|[\cos(t) + i\sin(t)]$, then one of the n nth roots of z is $|z|^{1/n}[\cos(t/n)+ i\sin(t/n)]$. Since the n nth roots are evenly spaced around a circle, they are separated by an angle of $2\pi / n$.

So, once we find the root $|z|^{1/n}[\cos(t/n)+i\sin(t/n)]$, we find the other n–1 roots by adding multiples of $2\pi / n$ to t/n.

Example: Find the cube roots of 1+ i.

First, write 1 + i in trigonometric form:

$$\sqrt{2}\left[\cos\left(\frac{\pi}{4}\right)+i\sin\left(\frac{\pi}{4}\right)\right]$$

Then one root is $\sqrt[6]{2}\left[\cos\left(\frac{\pi}{12}\right)+i\sin\left(\frac{\pi}{12}\right)\right]$. To find the two remaining cube roots, we add multiples of to to get and . Thus the three cube roots of 1 + i are

$$\sqrt[6]{2}\left[\cos\left(\frac{\pi}{12}\right)+i\sin\left(\frac{\pi}{12}\right)\right]$$

$$\sqrt[6]{2}\left[\cos\left(\frac{3\pi}{4}\right)+i\sin\left(\frac{3\pi}{4}\right)\right]$$

$$\sqrt[6]{2}\left[\cos\left(\frac{5\pi}{6}\right)+i\sin\left(\frac{5\pi}{6}\right)\right]$$

Exercise

Find the four fourth roots of –16 i.

Solution

First, we must write – 16 i in trigonometric form. The magnitude is 1, and the direction angle is π. So the trigonometric form of –16 i is 16 (cos π + i sin π).

There will be four fourth roots. The magnitude of each of the fourth roots will be the fourth root of the magnitude of –16 i. So the magnitude of each of the fourth roots will be 2. The angle of the first of the four fourth roots will be one-fourth of the angle for $-16i$, so the angle of the first of the four fourth roots will be $\pi/4$.

The four fourth roots will lie evenly spaced around the unit circle, so they will lie an angular distance of $\pi/2$ radians apart. So the four angles associated with the four fourth roots will be $\pi/4$, $3\pi/4$, $5\pi/4$ and $7\pi/4$. So the roots will be:

$2 [\cos (\pi/4) + i \sin (\pi/4)] = -2 + -2 i$

$2 [\cos (3\pi/4) + i \sin (3\pi/4)] = - -2 + -2 i$

$2 [\cos (5\pi/4) + i \sin (5\pi/4)] = - -2 - -2 i$

$2 [\cos (7\pi/4) + i \sin (7\pi/4)] = -2 - -2 i.$

Trigonometry and Vectors

A vector in the plane is a directed line segment (represented by an arrow) from some initial point P to some terminal point Q. Two vectors are considered equivalent if they have the *same length* and the *same direction.* When the initial and terminal points are named with letters such as P and Q, the vector can be represented as PQ with an arrow over the top, i.e. $\overrightarrow{PQ}$.

A *standard vector* is a vector in *standard position,* which means a vector with initial point at the origin in the Cartesian coordinate system. Every vector in the plane is equivalent to a standard vector.

Displacement is an example of a quantity measured by a vector. Suppose an object in a plane (the floor, for example) is moved three units to the east and four unit north. Then it would have been displaced a total of five units in an approximately north easterly direction along the diagonal of a right triangle with sides three and four.

In the science of Physics, force is a quantity measured by a vector. Every force has a magnitude and a direction of application. The 'length' of a force vector equals the magnitude of the force. Consider how one would 'add' two displacements. Suppose an object is moved 2 units west, then one unit south. Next, it is moved 3 units east and 4 units north.

The sum or result of the two displacements would be the same as if the object had been moved 1 unit east and 3 units north. Picture the first displacement as an arrow going from the initial position of the object to its next position. Then think of the second displacement as an arrow going from the second position of the object to the third position. Then the resulting displacement would be an arrow going from the first position of the object to the third position of the object. See the diagram:

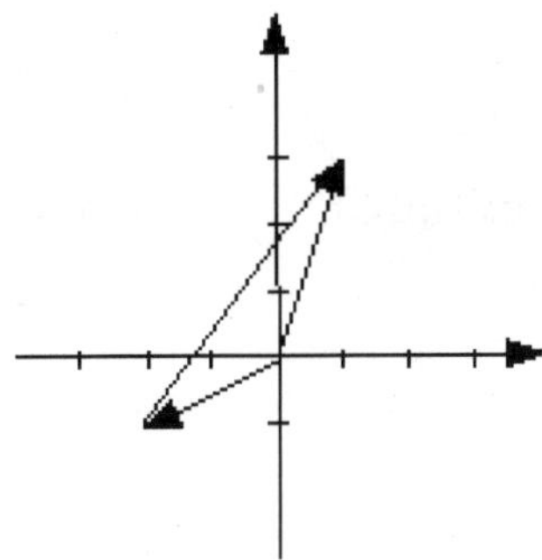

We see that the addition of vectors can be represented by placing the initial point of the second vector at the terminal point of the first vector, then the sum of the two vectors is the vector beginning at the initial point of the first vector and ending at the terminal point of the second vector. This is sometimes called the triangle rule for the addition of vectors.

Using this same example, we can introduce the idea of the *components* of a vector. Vector components are sometimes indicated by an ordered pair of numbers between two 'angle brackets'. Thus, the first displacement in this example—two units vest and one unit south—could be represented by the components $< -2, -1 >$.

The second displacement—three units east and four units north—could be represented by the components $< 3, 4 >$. Notice that it is easy to find the components of the sum or resultant vector—just add the components of the vectors. $< -2, -1 > + < 3, 4 > = < 1, 3 >$.

Now, consider a standard vector v = < a, b >. (We'll denote vectors with lower case bold letters.) The length or magnitude of of v is denoted by |v| and computed using the Pythagorean Theorem.

$$|\Delta| = \sqrt{a^2 + b^2}$$

Notice that $a/|v| = \sin\theta$ and $b/|v| = \cos\theta$, where θ is the angle between v and the positive x axis.

So $v = < a, b > = < |v| \sin\theta, |v| \cos t > = |v| < \sin\theta, \cos\theta >$. That is, $v = |v| < \sin\theta, \cos\theta >$.

The vector $< \sin \theta, \cos \theta >$ is called the *direction* of v.

Notice that we factored out the $|v|$ from the components of the vector. We can do this because multiplying a vector by a number changes the length of the vector accordingly. If we multiply a vector by 2, the length is doubled and the direction is unchanged.

If we multiply a vector by one-half, the length is halved and the direction is unchanged.

However, if we multiply by - 1, the length is unchanged and the direction is reversed.

In general, if $v = < a, b >$, the $c\,v = < c\,a, c\,b >$

Exercise

Find the magnitude of the direction vector $< \sin \theta, \cos \theta >$.

Solution

The magnitude of the direction vector $< \sin\theta, \cos\theta >$ equals the square root of the sum of the squares of the components. So the magnitude is 1. Direction vectors are sometimes called *unit* vectors because they have unit length.

Exercise

Given that $v = < -2, 2 >$, find the magnitude, the direction angle θ and the direction vector. Find the components of 3 v, and -2 v.

Solution

$v = < -2, 2 >, \; |v| = 2"2$

v is in quadrant II and $\tan \theta = -1$, so θ = 135Ú. The direction vector is $v/|v| = < -\sqrt{2}/2, -2/2 >$.

$3v = < -6, 6 >$ and $-2v = < 4, -4 >$.

Two special unit vectors are sometimes defined:

$i = < 1, 0 >$ corresponding to the direction angle 0° and

$j = < 0, 1 >$ corresponding to the direction angle 90°.

Every vector in the plane can be represented as a linear combination of the unit vectors i and j.

$< a, b > = < a, 0 > + < 0, b > = a < 1, 0 > + b < 0, 1 > = a\,i + b\,j$

Thus, $a\,i + b\,j$ is an alternate way of writing a vector in component form.

Exercise

Find the components of a vector of magnitude 4 and direction angle 45°.

Solution

$$|v| = 4, \theta = 45^\circ$$

$$v = |v| < \cos 45^\circ, \sin 45^\circ > = \left\langle 2\sqrt{2}, 2\sqrt{2} \right\rangle$$

Sometimes vectors are represented solely in terms of their magnitudes and directions, with specifying their components.

Consider the following problem:

Two vectors u and v of magnitude of 5 and 2 respectively are separated by an angle of 30°. Find the magnitude of their sum u + v and the angle the sum makes with the first vector.

The following is a geometrical diagram of the problem:

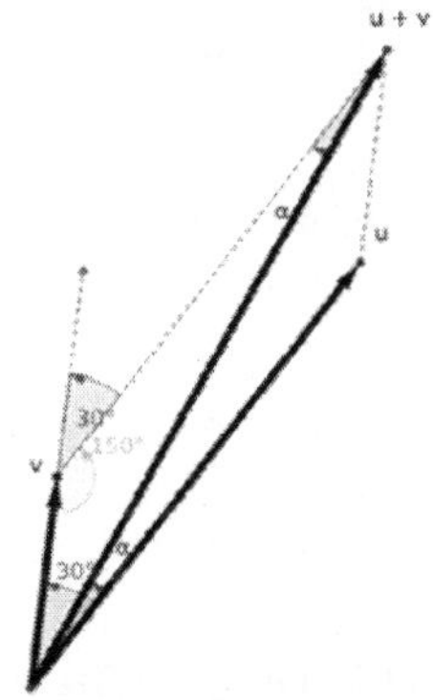

The vectors u and v are initially drawn from the same initial point so that we can show the 30° separation. A copy of u is then moved so that it begins at the terminal point of v. This new copy of u makes an angle of 180° – 30° = 120° with v. This gives us a triangle with sides 2 and 5 and an included angle of 120°. Thus, we can find the magnitude of u + v using the Law of Cosines.

$|u + v|^2 = 5^2 + 2^2 - 2\ (5)\ (2) \cos 150^\circ = 29 + 20 \cos 30^\circ = 46.3205$ approximately.

So $|u + v|$ is approximately 6.806.

To find the angle θ the sum makes with u, note that it is congruent to the angle opposite v in the triangle. Thus using the Law of Cosines again,

$\cos\theta = 0.9891$. Thus, the angle is approximately 8.45°.

Exercise

Two vectors u and v of magnitude of 3 and 2 respectively are separated by an angle of 45°. Find the magnitude of their sum u+v and the angle the sum makes with the first vector.

Solution

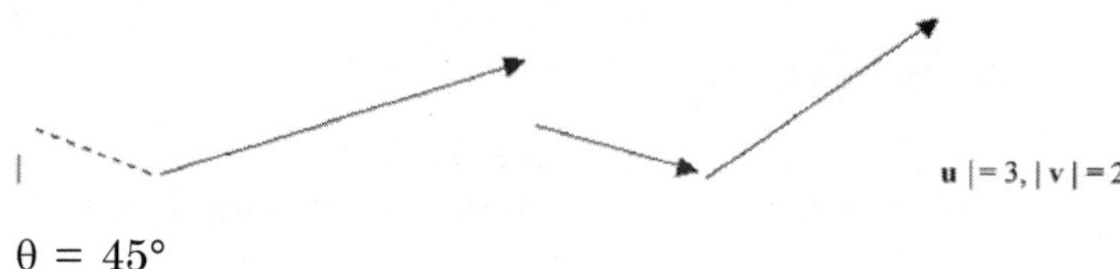

$\theta = 45°$

$180° - 45° = 135°$ is the angle of the triangle opposite the side representing the sum of the vectors.

The angle opposite $|v|$ is the angle between the vector u and the vector u + v.

$$|u + v|^2 = |u|^2 + |v|^2 - 2|u|\,|v|\cos 135° = 13+6\sqrt{2}$$

So, $|u + v| = 4.635$

Let ϕ denote the angle between u and u + v. Then using the Law of Cosines

$$\cos\varphi = \frac{|u|^2 + |u+v|^2 - |v|^2}{2|u||u+v|} \approx 0.9523$$

$$\varphi \approx 17.8°$$

An Ontology for Engineering Mathematics

Introduction

Engineers use mathematical models, such as sets of equations, to analyse the behaviour of physical systems. The conventional notations for formatting mathematical expressions in textbooks and in the engineering literature usually leave implicit many of the details required to understand the equations.

For instance, it is not clear from the expression $f = kx+c$ which symbols are variables or constants; whether they represent numbers or physical quantities (e.g., forces, lengths); whether the magnitudes are reals, vectors, or higher-order tensors; whether the quantities are static values, functions of time, or functions of time and space; and how units of measure are treated. The reader must *interpret* these notations using background knowledge and context. This is error-prone for humans and beyond the capability of today's computer agents.

To enable the sharing and reuse of engineering models among engineering tools and their users, it is important to specify a conceptual foundation that makes these distinctions explicit and provides a context- and reader-independent semantics. Toward this end, we have developed a formal ontology for mathematical modelling in engineering, called EngMath. The ontology builds on abstract algebra and measurement theory, adapted to meet the expressive needs of engineering modelling. The specification includes a first-order axiomatization of representational vocabulary that is machine and human readable.

This paper is about the EngMath ontology, and how it exemplifies the design and use of such ontologies in support of agent communication and knowledge reuse. Such an ontology differs from what is found in engineering textbooks and philosophy books in that it is designed as a specification for these knowledge sharing purposes. We begin in Section 2 by describing the role of ontologies as formal specification and the uses of the EngMath ontology.

In Section 3, we give define the basic concepts and relations in the ontology. In Section 4, we discuss a series of design decisions and their rationale. In Section 5, we offer design criteria—minimizing ontological commitment and maximizing monotonic extendibility—and demonstrate techniques used to achieve them. In Section 6, we discuss the relationship of the EngMath ontologies to relevant work in philosophy and AI

The Purpose of the Ontology

Ontology as Formal Specification

A body of formally represented knowledge is based on a *conceptualization*: the objects, concepts, and other entities that are assumed to exist in some area of interest and the relationships that hold among them. A conceptualization is an abstract, simplified view of the world that we wish to represent for some purpose. Every knowledge base, knowledge-based system, or knowledge-level agent is committed to some conceptualization, explicitly or implicitly.

For the purpose of knowledge sharing, formal ontologies serve as *specifications of common conceptualizations* among agents. In the philosophy literature, ontology is the systematic account of Existence—aiming to account for all forms and modes of being. For AI systems, what can exist in a conceptualized world is determined by what can be represented. If agents are to communicate in a shared language or if a body of formally represented knowledge is to be reused, then there must be some agreement about a universe of discourse. Furthermore, if the shared language includes vocabulary denoting entities and relationships in the conceptualization, there must be some way to specify what can be meaningfully stated in this vocabulary. Ontologies, in the context of knowledge sharing, are a means for making such content-specific agreements.

If we assume a common syntax and semantics for a core representation language, then we can specify conceptualizations by writing definitions of shared vocabulary. That is the strategy proposed by the ARPA Knowledge Sharing Effort, and is the tack we are taking. A Knowledge Interchange Format (KIF) serves as the language for making assertions and definitions, and ontologies provide axiomatic and textual definitions of relations, functions, and objects. By 'definitions' we mean specifications of the well formed use of the vocabulary. Definitions include axioms that constrain the interpretation. Such an axiomatization specifies a logical theory, but is not intended as a knowledge base. Instead, the ontology serves as a domain-specific representation language in which knowledge is shared and communicated.

In practice, our ontologies define the vocabulary with which queries and assertions are exchanged among interoperating agents, some of which may be passive. The agents conform to ontological commitments which are agreements to use the shared vocabulary in a coherent and consistent manner. An ontological commitment is a guarantee of consistency, but not completeness, with respect to queries and assertions using the vocabulary defined in the ontology. Committed agents may "know" things not implied by the shared ontologies, and may not be able to answer queries that follow from the shared ontologies. Furthermore, the "shared knowledge" of these agents can be viewed at the Knowledge Level, as attributed and independent of symbol-level encoding. Thus, the agents may operate on any internal representation desired, as long as they use the shared vocabulary consistently in communication. This model of agent collaboration is being pursued by several groups.

Uses of the EngMath Ontology

In designing the EngMath ontology, we anticipate three kinds of use, and accept them as requirements. First, the ontology should provide a machine- and human-readable notation for representing the models and domain theories found in the engineering literature. Second, it should provide a formal specification of a shared conceptualization and vocabulary for a community of interoperating software agents in engineering domains. Third, it should provide a foundation for other formalization efforts, including more comprehensive ontologies for engineering and domain-specific languages. In this section we will give examples of each application.

EngMath as a Shared Notation

Engineers use mathematical expressions, such as constraint equations, to describe, analyse, and communicate models of physical devices and their behaviour. The quantities represented in these expressions are different from purely numerical values, and the algebra for operating over them must account for extra-numerical considerations such as dimensional consistency, units of measure, and vector and tensor operations. Some form of this 'physical algebra' is taught in nearly every introductory physics or engineering course, and the subject is prominent in the standard texts. Students are taught to use dimensional consistency to check for modelling and equation solving errors. A technique called Dimensional Analysis is used in design and to assist in the interpretation of experiments, and is an important area of study itself.

Textbook notations for physical quantities vary by author and leave much implicit—relying on context and background knowledge of the reader for proper interpretation. The problem of implicit notation is revealed when students try to encode engineering models using mathematical support software. Human expertise is required to map expressions about physical quantities to the purely mathematical constructs of current commercial math tools.

The EngMath ontology is intended to provide a formal language sufficient to express the models in engineering textbooks and to map them to mathematical software tools. We view the latter application as an instance of agent communication, which is the subject of the next section.

EngMath as a Vocabulary for Agent Communication

By providing a declarative, machine readable representation, the EngMath ontologies can enable unambiguous communication between software agents that would otherwise be difficult or impossible.

To illustrate a use of the ontology, consider a simple example of agents exchanging symbolic representations of spring behaviour. Agent A is a specialist in the design of springs, and agent B is a specialist in quantity algebra. Agent A needs a solution to a set of equations relating spring and material properties that include the following:

$$k=\frac{d^4 G}{8D^3 N}, \; G=11{,}5000\,\text{kpsi}$$

where *k* is the spring rate, *d* is wire diameter, *D* is spring diameter, *N* is number of turns, and *G* is the shear modulus of elasticity.

Agent A can send Agent B these equations as a set of KIF sentences, using the vocabulary of the EngMath ontology:

```
(scalar-quantity k)
(= (physical.dimension k)
  (/force-dimension length-dimension))
(scalar-quantity d)
(= (physical.dimension d) length-dimension)
(scalar-quantity dm)
(= (physical.dimension Dm) length-dimension)
(scalar-quantity N)
(= (physical.dimension N) identity-dimension)
(scalar-quantity G)
(= (physical.dimension G)
  (* force-dimension
   (expt length-dimension -2)))
(= k (/(* (expt d 4) G) (* 8 (expt Dm 3) N)))
(= G (* 11.5 (expt 10 6) psi))
```

After receiving the equations in this form, agent B can answer questions about the values of the terms such as the diameter (d). The vocabulary used in this interaction, such as the function constant physical.dimension, is independent of a domain theory for springs. The sentence simply states an algebraic relationship between quantities and the dimensional characteristicsof those quantities.

This type of information allows Agent B to perform algebraic manipulations such as solutions of simultaneous equations or numerical evaluations of individual parameters. Because dimensional information is included, the consistency of equations can be checked by the agent and agents are freed from committing to an explicit set of units.

The spring example is typical of problems in introductory engineering textbooks. More complex interactions are required to coordinate commercial design tools on industrial problems.

In the SHADE project, we have constructed a set of software agents that interact to support collaboration on industrial problems like satellite system design. One SHADE agent is a specialist in rigid body dynamics (RBD), and another is responsible for the geometric layout of satellite components (Layout). Both commit to the EngMath ontology.

The RBD agent queries the Layout agent about the inertial characteristics of a particular component. These characteristics include the mass whose value is a scalar quantity and the inertia tensor whose value is a second order tensor.

In reply, the Layout agent specifies the inertia tensor with respect to a global reference frame and point. The reference frame is part of the shared domain theory for the two agents; it is not implicit in the representation of the tensor. This allows the RBD agent to translate the inertia into a different reference frame convenient for dynamic analysis.

Most SHADE agents are commercial tools wrapped so that they conform to ontological commitments and communication protocols. These agents are designed to be conformant at the interface, but are not required to represent the ontologies internally. Some agents can assimilate an ontology and use it as input. The Unit Conversion Agent is an example. Its contract is specified entirely by the EngMath ontology. This agent takes KIF expressions over quantities, and performs services such as symbolic simplification, unit conversion, and dimensional consistency verification. It can *read* ontologies that specify of other unit systems, and determine whether the system is complete for the dimensions specified.

EngMath as a Conceptual Foundation

The EngMath ontology was also designed as a conceptual foundation for other ontologies; in its design we needed to anticipate how it would be used in other theories. For example, we are constructing a family of ontologies for representing components and relations among them (e.g., part-subpart relations, connections, association of component features and constraints).

One ontology of mechanical components, for instance, is for models in which components have mass properties and associated reference frames and points, but lack complete geometric representations. This theory combines an abstract component ontology, a constraint expression ontology, parts of the EngMath ontology, and a simple geometry theory (that also includes EngMath).

The Compositional Modelling Language (CML) is another example of building on the EngMath ontologies. CML is a modelling language that is intended to synthesize and redesign the various formulations of Compositional Modelling to enable model sharing among research groups. Part of the language design of CML is an ontology about time, continuity, object properties, etc.

The semantics of the language are specified axiomatically, using the vocabulary of the CML ontology. The CML ontology builds on the EngMath ontology as a foundation.

Algebraic Properties of Quantities

We can borrow the well-established theories of algebra for the reals, vectors, and higher-order tensors. To adapt them to physical quantities we must consider physical dimensions in describing the domain and range of operators. We have already seen how dimensional homogeneity is a prerequisite for adding quantities, and physical dimensions establish classes of quantities that are comparable. For each physical dimension, the class of constant scalar quantities of a physical dimension forms an abelian group with the addition operator + and a zero identity element for that dimension (zeros of each dimension are different). The class of all scalars of any dimension, after removing the zero scalars, forms an abelian group with respect to multiplication.

For vector quantities, the sum of the quantities is only defined where the sum of the dimensionless versions of the vectors would be defined (i.e., the spatial dimensions must align). For higher-order tensors, tensor order and spatial dimensions, as well as the physical dimension, must be homogeneous. Analogous restrictions apply for the multiplication of tensors. For function quantities, the sum or product of two function quantities is another function quantity that is only defined where the domains of the functions are equal. For unary scalar function quantities, the addition and multiplication operators are defined to handle a mix of function quantities and constant quantities.

The sum of a constant k and a time-dependent function f, for example, is a function defined everywhere $f(t)$ is defined and is equal to $f(t)+k$. Continuous time-dependent quantities can also be defined from others using a time-derivative function. Most of the axiomatization in the specialized theories for functions, vectors, and tensors is concerned with specifying the conditions under which algebraic operators apply. This is essential for building agents with guarantees of completeness for some class of quantities.

Rationale for Important Distinctions

There are many possible ways to axiomatize this domain, and the axiomatization makes explicit many things that are implicit in the engineering literature. In this section, we discuss important distinctions that the formalization process forces one to clarify. For each distinction we offer a rationale for the choices made in terms of the purpose of the ontology.

Quantity Types Versus Instances

Some authors define "a quantity" as a set of values or a property having some order. However, they offer no name for instances of this set, or for the values of this property. Following the AI convention of defining concepts as classes (or equivalently, sets, types, monadic predicates), we define the class physical quantity, and define species of quantities as subclasses of physical quantity.

Where some would talk of the unitary concept "the mass quantity," we would say that there is a quantity type whose elements are constant scalars of dimension mass. This allows us to support the common usage in engineering modelling, where specific values such as the length of a beam are called quantities. It also allows us to modularly state properties of quantities, function quantities, scalar quantities, scalar-mass quantities, etc., and have the general properties inherit to the specializations.

Quantities are not Variables

Physical quantities and physical dimensions are objects in the universe of discourse, and are not linguistic elements (e.g., "variables" of constraint expressions). Constraints over quantities that are found in engineering models are typically algebraic equations or inequalities, sometimes including the operations of differential calculus. Quantities are very different from linguistic elements. We have built ontologies (for the specification of a configuration design task for elevators—the VT experiment) in which constraint expressions and their constituents (variables and arithmetic operators) are part of the domain of discourse.

In that formalization of constraints, constraint expressions are logical sentences with nonlogical constants (not logical variables) denoting physical quantities. The ontological distinction between variables and quantities allowed us to write an ontology in which both the form and denotation of constraint expressions were specified. This is necessary when the committing parties need restrictions on the form of expressions to guarantee completeness.

Quantities are not Properties of Objects

The identity of quantities is independent of physical objects that might be modelled. For example, let us say that the mass of a body B is the quantity 50kg. 50kg is just a measure of mass; there is nothing intrinsic about 50kg that says it is the mass of B (there is no total function from masses to bodies). Making quantities independent of objects also allows one to state that the mass of B is *equal* to the mass of a different body. It also supports pure parametric models, in which there are no physical objects and only quantities are described.

Of course, the model builder is free to define a function from physical objects to quantities, but this function is not the same thing as a mass quantity. Our formulation of quantities does not preclude such object-to quantity functions; it provides the language in which to describe the range of those functions.

In CML, for example, there are functions from objects (e.g. pipes) to time-dependent function quantities. This distinction is central to the semantics of CML, which allows both time-varying and static relations. The object-to quantity functions are time-independent (e.g., the pipe always has a flow rate), but the quantities are functions of time (e.g., the flow rate is a function quantity that has different values throughout a scenario and can be undefined on some values of time).

Quantities are not Tuples

In our ontology, quantities exist as values that are related to, but independent of units and numbers. Quantities can be compared, multiplied, etc., without ever converting to reals or considering units of measure. This independence allows the model builder the flexibility needed to build measurement systems and physical theory independently.

Furthermore, although quantities have algebraic properties, they are not purely abstract entities; we have grounded them in the modelled world with the condition of comparability. In contrast, an alternative formulation one often sees is to treat quantities as tuples of numbers and units or as simply numbers.

Besides the assumption about units that such a formulation makes, we find that it contradicts the conceptualization of quantities in the world made by physicists and philosophers. To reuse a simple example: 3 feet and 1 yard are equal, yet the tuples <3,ft> and <1,yd> are not.

There are no Fundamental Quantities

The identity of a physical quantity is also independent of any fundamental quantities. We make no ontological distinction between base and derived quantities. For any pair of comparable physical quantities, their sum and product (if defined) are also physical quantities (the sum will be comparable with the original two; the product may not).

The rationale for this decision is again to provide generality and flexibility. We know from physics that there is no physical basis for giving some quantities primacy. Measurement theories make a distinction between quantities amenable to fundamental and associative measurement. Again, we have made this distinction irrelevant for engineering models by avoiding the temptation to define quantities in terms of measurement. The analogous argument holds for not giving special status to some units of measure or physical dimensions. Even though it is possible to *compose* units and dimensions, there is no need to stipulate in the shared theory exactly which units and dimensions are fundamental.

Nonphysical vs. Dimensionless Quantities

The conceptualization leaves it to the model builder to define the fundamental dimensions for a domain. The definition of physical dimension does not preclude one from defining new dimensions not mentioned in physics texts. We cited the example of amount of money as a possible physical dimension.

The ontology also allows for dimensionless quantities, such as real numbers, whose physical dimension is the identity dimension. How does one decide whether to define a new physical dimension for a nonphysical quantity or to make it a dimensionless quantity? We say that a physical dimension distinguishes a type or class of quantities that can be meaningfully combined with algebraic operations, can undergo unit conversion, and are comparable.

Amount of money is a meaningful dimension because one can accumulate sums of money, do currency conversion, and compare relative wealth. Amount of money can be meaningfully combined with other dimensions.

A rate of inflation, for example, is computed by dividing an amount of money (a change in price) by a different amount of money (the base price) and dividing the result by a unit of time (a year). The rate is a quantity of dimension (expt time –1). Money is something to be tagged as different in type from other quantities; that's why we see dollar signs (or other units) carried with the numbers in formulae.

As a negative example, consider quantities like number of politicians and number of constituents. We might write a formulae

(= Nconstituents (* Npoliticians 1000000))

In this model, we are making an abstraction; these quantities of humans are being compared as numbers. The = sign implies that the quantities must be of the same physical dimension, which would be, in this case, the identity dimension. Suppose, however, that we wanted to describe the number of molecules in those politicians.

There is an international unit for number of molecules, the Mole. In the modern SI system of units, this is not a dimensionless quantity, but a quantity of the dimension amount of substance. Why does amount of substance get special status over number of politicians? Because chemical models need to distinguish amount of substance quantities from other measures.

There is something homogeneous about molecules in that it makes sense to measure stuff by counting them. The formula for the average amount of gas in those politicians would use an amount of substance quantity for the amount of gas, and a dimensionless quantity for the number of politicians. This makes sense in the chemist's abstraction of politicians: that they can be viewed as volumes of gas.

Measurement Scales

A measurement scale is a determinative, non-degenerative assignment of numeric values to physical quantities. *Determinative* means that the same quantities are consistently assigned the same numeric values, and *non-degenerative* means that different quantities get different values (ignoring issues of precision).

Examples of measurement scales include Mohs scale for the hardness of minerals and the Celsius and Kelvin scales for thermodynamic temperature.

Using the Coombs classification, the Mohs scale is *ordinal* because it only provides information for inequality comparisons among points on the scale. The Celsius scale is classified as *ordinal-interval* because it supports inequalities between intervals on the scale, and Kelvin is a *ratio* scale because it supports comparisons of the ratio of any two values on the scale. For instance, it makes sense to say that 6 degrees Kelvin is twice as much as 3 degrees Kelvin.

By this mathematical definition, the units of measure in our conceptualization correspond to ratio scales.Each value on the scale is the ratio of the measured quantity to the degree-Kelvin unit. Thus the Kelvin scale can be defined from the degree-Kelvin using the magnitude function:

(lambda (?q) (magnitude ?q degree-kelvin))

We don't call Mohs ratings or degrees-Celsius units of measure, because they aren't quantities against which to compare other quantities. Of course one can write a function that does measure any temperature on the Celsius scale, such as

(lambda (?q)

(–(magnitude ?q degree-kelvin) 273.15))

Since there is no principled constraint on the form of such function, we leave it to the model builder to define scales appropriate to a domain.

Design and Evaluation Issues

We view ontologies as designed artifacts, formulated for specific purposes and evaluated against design criteria. In a separate paper, we propose a set of ontology evaluation criteria and show examples of their application to this and other domains.

Two of the criteria will be illustrated here: to minimize *ontological commitment* while allowing for *monotonic extendibility*. Minimizing ontological commitment means making as few claims as possible about the world being modelled, allowing the parties committed to the ontology freedom to specialize and instantiate the ontology as needed. Extendibility means an ontology should be crafted so that one can extend and specialize the ontology *monotonically*.

In other words, one should be able to define new terms for special uses based on the existing vocabulary, in a way that does not require the revision of the existing definitions. Both of these criteria hold a natural tension with the goal of supporting the knowledge sharing needs of a range of agents with differing abilities and assumptions.

Adding vocabulary to handle a broad range of representation needs will increase ontological commitment by adding more constraints on the interpretation, and will make it more likely that some definitions will be incompatible with future representation needs.

In this section, we will discuss two techniques in the design of the EngMath ontology that help us meet these two criteria. One is the decomposition of a large ontology into modules. The second is another familiar design technique—parameterization—applied to the problem of representing conventions.

Decomposing the Ontology

The first technique is to decompose a monolithic ontology into a set of loosely coupled sub-ontologies. The EngMath ontology is decomposed into an inclusion lattice of individual ontologies, where each ontology is a set of definitions, and ontology inclusion is set inclusion. If a ontology B *includes* ontology A, then ontology B is the union of the definitions in A with those specific to B. More sophisticated methods for partitioning knowledge bases into modular theories are being explored, but set inclusion is sufficient for mutually consistent ontologies in a uniform namespace.

The inclusion lattice of theories for the EngMath family. The core ontology of physical quantities includes abstract algebra (evolved from the example in the KIF 3.0 specification) and a theory of objects and relations called the Frame Ontology. The EngMath family includes ontologies for Scalar Quantities, Vector Quantities, and Unary Scalar

Functions. The Standard Units ontology defines many of the most common physical dimensions and units, and includes the SI system. Other engineering ontologies that build on the EngMath family—for describing component structure, design tasks, discrete events, and specific analysis domains such as kinematics—are being developed.

Decomposing into loosely coupled ontologies helps minimize ontological commitment by allowing one to commit to a coherent subset of the axioms of the entire ontology.

For example, one can commit to scalars but not vectors; unary functions (for ODE models) but not n-ary functions (for PDE models); and static models (no functions). Even an agent that does not include physical dimensions in its conceptualization can be accommodated, since all the realnumber operations are also defined for quantities, and the reals are quantities.

Decomposition also supports the specialization of monotonically extendible ontologies. For example, the CML ontology inherits the basic commitments of the Unary Scalar Functions ontology and adds notions specific to its needs. Since the definitions in Unary Scalar Functions anticipate the general class of time-dependent functions, the CML extensions were possible without redefining the originals.

Designing a family of coherent ontologies is more difficult than designing a monolithic one, since the designer must anticipate intermodule interactions, such as axioms in several ontologies constraining the same vocabulary. To help manage the extension of vocabulary across ontologies, we borrow a technique from software engineering: *polymorphism*.

Polymorphism allows the same function constant to be defined in several places, where each definition adds axioms about the use of the constant.

The +, *, and expt functions, for example, are polymorphically extended to each type of quantity and to physical dimensions. The form of a polymorphic definition of a function F is

```
(=> (and (p ?x) (q ?x))
(=> (= (F ?x ?y) ?z) (r ?x ?y ?z)))
```

where ?x and ?y are the arguments of F, p and q are the domains over which this definition applies (e.g., quantity types), and r is the condition that holds for F's arguments and it's value, ?z.

Due to theory inclusion, the definition of a function is the union of axioms contributed by each theory. For example, the definition of + for the Vector Quantities ontology is the union of the axioms for + for vector quantities, + for scalar quantities, + for physical quantities in general, and + for the reals.

Parameterizing Conventions

A second technique to facilitate extendibility and to minimize commitment is the parameterization of conventions. A parameter of a black box system is a representation of assumptions about how the environment interacting with the system will vary.

In a computer program, the environment interacts with the program through formal arguments that are bound on invocation (and through the human interface, which is not as neatly parameterized). In the domain of the EngMath ontologies, the environment is the instantiation of the shared vocabulary for particular domain models. We have described several ways in which the domain models vary *by convention*. The choice of fundamental physical dimensions and standard units for a system, and the choice of particular units for individual quantities are conventions.

To minimize ontological commitment, we formulate these choices as parameters of the engineering model, rather than global constants of the shared ontology. To support extendibility, we provide an expressive representational vocabulary for specifying values of parameters. For example, to allow the model builder to specify the choices of fundamental dimensions and standard units, we provide the machinery to define systems of units.

It is in this sense that ontologies are a "coupling mechanism" for knowledge bases and knowledge based agents. Parameters such as the system of units for a domain model are like abstract data types in software, except that the latter are ground and parameters of a domain model can be *theories*. We mentioned that the Unit Conversion agent can take, as inputs, ontologies specifying systems of units. When the other SHADE tools exchange sets of equations they are also exchanging theories. When agents pass around theories, the constraints on what they can pass around is specified in a shared ontology. In this sense, ontologies play a similar role as database schemata, except that ontologies may require a more expressive language for constraints than is typical in database specifications.

Related Work

Philosophy

The mathematics underlying the EngMath ontologies is drawn from standard textbooks used to teach engineering math, and from the philosophy literature on measurement. We had to choose distinctions from the conceptualizations, and modify them to produce the modular, internally coherent ontologies in the EngMath family. The engineering texts assume a conceptualization motivated by abstract algebraic operations, and the philosophy texts explore the grounding of the abstractions.

The aim of the philosophy texts is to describe The World as it Is in its Entirety, and to relate the results to prior writings. For the philosopher, it is important to relate the nature of quantities to the process of measurement and observation, and more generally, to the question of scientific knowledge.

For instance, even the very notion that quantities such as mass can be meaningfully expressed as linear ratios of a standard turns out to be a convention, albeit a very useful and familiar one. Ellis argues that the notion of a unit is incomplete without a choice of how such units are combined (which is related to, but not the same as, measurement procedure). We are assuming that there is a shared interpretation to the result of multiplying the meter times the real number 1000.

For our purposes—the sharing and communication of engineering models in machine and human readable forms—it is an *advantage* to be able to isolate the meaning of quantities from the process of measurement. We make no apologies: this is not a sweeping-under-the-rug of relevant issues, but a strategic decoupling of issues. In building ontologies we are writing social contracts.

We are free to invent the conceptualization as long as its meaning can be effectively communicated (which is why we use standard terminology and logic). By accepting the KIF language as a foundation, for example, we already commit to including abstract things like sets and relations in the universe of discourse. We add to that an ontology of abstract algebra, and extend it to physical quantities. The philosophical ontologist draws a heavy line when the objects cross from timeless mathematical entities to physical concepts like mass and length. Our agents need no strong boundary; if our engineering model states that a quantity of mass exists and is related to a length quantity by some algebraic expression, it is so for the agent. According to our evaluation criteria, clarity and coherence in the specification are paramount, and faithfulness with The World is not an issue.

Nonetheless, we can "share and reuse" the analysis found in philosophy writing, for very pragmatic ends. For example, one of the differences between a casual writing on quantities and a careful one is the treatment of property association. One often sees it stated that quantities are "properties" of "objects" — like qualities but quantitative.

However, the careful writer will point out that quantities as we use them are *relational*—they are about comparing objects in some respect, or about relationships among them. This informs our design. Departing from the conventional "object oriented" approach, we give independent status to quantities and leave it as a modelling decision to map from objects to quantities. This is, in essence, a decoupling of theories about quantities from theories about model formulation.

Similarly, an understanding of the nature of physical theory and measurement guides us away from the impulse to oversimplify for the sake of computational elegance. For example, it would be simpler from a computational point of view to fix a single basis set of "fundamental" physical dimensions or units of measure, and recursively derive the rest.

However, the laws of physics tell us that there are no inherently privileged physical dimensions, and the study of measurement tells us that the choice of basis sets for dimensions and units is a convention. The fact that engineers must deal with at least two systems of units, each of which chooses a different basis set (and which has changed over historical time), motivates us to provide the model builder with the representational machinery to define a system of units as part of the domain model.

Other Work on Ontologies

Work in Formal Ontology in the philosophy literature that is relevant to knowledge sharing ontologies are found in, and (combined with papers from AI).

There is a growing body of ontologies appearing in the literature seen by the knowledge representation community, including as a sample.

The most closely related on engineering ontology is the thesis by Alberts. Alberts describes a formal ontology intended as the basis for building interoperable and reusable knowledge systems for design. His ontology provides a vocabulary for modelling the structure and behaviour of systems, based on systems theory and finite element modelling.

While the formalization of systems is exemplary, the treatment of quantities in that ontology is simplistic. First, quantities have no status other than as the values of variables with symbolic 'quantity types.' There is no provision for defining new dimensions or describing complex dimensions; the quantity types appear to be those that are anticipated by systems theory (analogs of effort and flow).

Second, the values are always unary scalar functions of time, where time is "exceptional" (i.e., outside the range of the model). This prevents vector and tensor models, PDE models, phase-space models where time is not the independent variable, etc. Third, the values of these functions are tuples of numbers and fixed units.

More recent work by Akkermans and Top develops the systems theory methodology to maturity. It proposes engineering ontologies at four levels of description: functional components, physical processes, mathematical relations, and model data.

Work on model formulation using the CML language and SHADE engineering agents aims at a suite of ontologies not based on system theory that have similar coverage. Perhaps the ontology presented in this paper can provide a foundation for the mathematical and data level of models in these comprehensive engineering ontologies.

Functions and their Graphs

Functions, Domain and Range, Graphs of Functions

So what exactly IS a function? A mathematical definition would be..

Definition: FunctionA relationship in a mathematical equation where one value of "x" yeilds one and only one value of y.

So what does that mean? Well to put it simply. If we have an equation, and we put a value of "x" in it and evaluate for "y", we will only get ONE answer for "y".

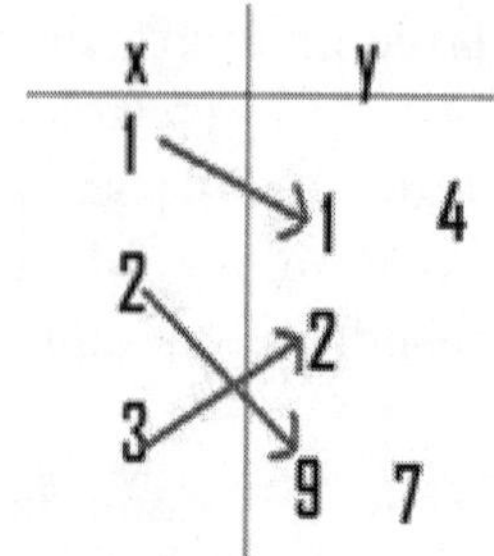

This relationship DOES represent a function. Notice how only one value for “x’ is going to one value of “y”. 1 goes to 1, 2 goes to 9 and 3 goes to 2. Also notice that in the “y” section, ther are other numbers that aren’t used. That’s okay, you can have numbers not used in that section. However you CAN NOT have numbers without a “partner” in the “x” section.

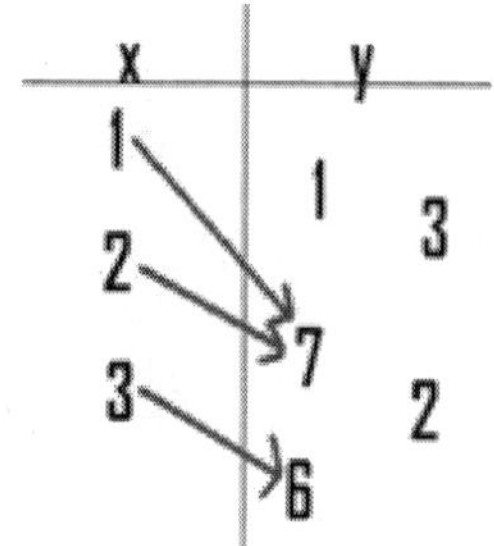

This also represents a function. Just because a number in the “x” column goes to two numbers in the “y” column, doesn’t mean it isn’t a function. It still follows the definition. One value of “x” is going to one and only one value of “y”.

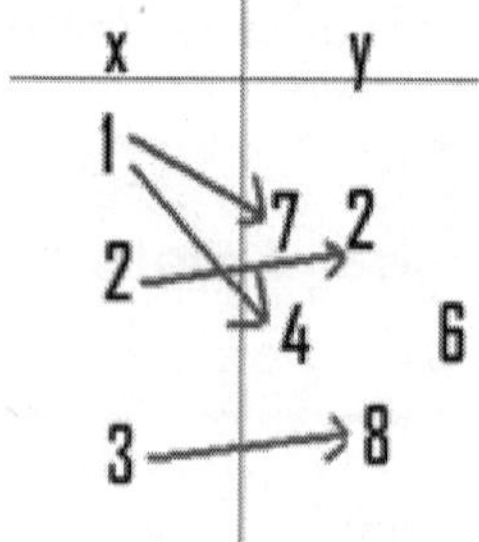

This does NOT represent a function. Notice that the number “1” has two arrows coming out of it. This means when we plug “1” in for “x” we get two numbers (solutions). Therefore we do not have a function.

Keep in mind that a function isn't only restricted to an "x" and "y" relationship. It could, for example, be a "distance" and "time" relationship. "d" and "t". All the same rules still apply however. Lets apply this concept to a mathematical equation.

$$y = 2x + 1$$

In this expression, if we plug in a value for "x" we will get one and only one value for "y".

So what exactly is this concept of plugging things in and getting one value? Well we call this the domain and range of the function. The domain would be the "x" demonstrated in the previous examples and the range would be the "y" in the previous examples. Here are more formal definitions for Domain and Range.

So in the above expression, is there any value of "x" that will make the value of "y" not true? No there isnt. So we can say the domain of the function is all real numbers. Which means, the acceptable replacements for "x" on that function is any number on the real numberline.

So then what would be range be? Well since we can put all real numbers IN TO the function, the solutions will be all real numbers. Which means this function exists (or has solutions on) on all numbers up to infinity.

This simply states that the "y" in a function can be replaced with "f(x)" But what does "f(x)" mean anyway? Well the "f" in "f(x)" is the name of the function. That could be anything you would like. For example, "g(x) h(x) Bill(x)" ect..

The three most common names for functions are "f" "g" and "h". The "(x)" in the notation means, in the function of "f" replace the variable with "x" Lets go through an example of this notation in a problem. Lets use the same function we used before only lets write it in function notation.

$$f(x)=2x+1$$

Now if we replace the "x" in this problem with a number, that means in the function "f" replace "x" with the number and evaluate. So if we plugged in a 2 we would have this.

$$f(2) = 2(2)+1$$

$$f(2) = 4+1$$

$$f(2) = 5$$

This notation can be useful if we want to subtract the solutions of the same function.

For example, if we wanted to subtract the solution we get from plugging in "10" for "x" and the solution we get from plugging "2" in for "x", that would be denoted as such.

$$f(x)=2x+1$$

$$f(10)-f(2)$$

$$[2(10)+1]-[2(2)+1]$$

$$21-5$$

$$=16$$

This concept doesn't just pertain to arithmetic, but it also works for multiplication and division.

In this next example we will have two functions. Each has a different name (f(x) and g(x)). The operations would go like this.

$$f(x)=2x+1$$

$$g(x)=4x-3$$

$$f(x)g(x)=(2x+1)(4x-3)$$

$$=8x^2-2x-3$$

$$\frac{f(x)}{g(x)}=\frac{2x+1}{4x-3}$$

Since we are only doing "f(x)" and "g(x)" we can just leave the answer as the expression and not solve for "x". Same thing goes for the division one, there isn't and further simplification to be done.

There are other functional operations we can do. Now that we have defined functions and what makes up a function, lets move on and look at the different kinds of graphs that functions can create. But first..

Properties of Lines

When you think in the concept of "slope" or steepness of the line, what would a horizonal line be?

The slope of a horizontal line is always zero. Because you are rising zero and running "x" (Where x is equal to any number.) But what about a vertical line? You are rising "x" and running zero. X over zero. This is UNDEFINED! You can NOT have a zero in the denominator in mathematics. It just can't be done. So when a line is vertical we say it is undefined.

Parallel Lines

A parallel line is a line that has the exact same slope as another but a different y-intercept.

To find a parallel line that goes through a given point, use the same slope, and just plug in the given point and solve for b. Here is an example.

Find a line parallel to the line you just calculated that passes through the point (–4,3).

So we will take the equation of the line we just calculated.

$$y=\frac{3}{2}x-\frac{1}{2}$$

$$y=\frac{3}{2}x-b$$

$$3=\frac{3}{2}(-4)-b$$

$$3=-6-b$$

$$-b=9$$

$$b=-9$$

$$\boxed{y=\frac{3}{2}x-9}$$

Perpendicular Lines

A perpendicular line is a line that passes through another line at a right angle.

To find this line, you take the slope of the line and take the negative reciprocal of it. Then plug in a given point and solve for b. Here is an example of this.

We are going to start with the first equation we calculated.

Find the line perpendicular to the line we just calculated and passes through the point (–4,6).

$$y = \frac{3}{2}x - \frac{1}{2}$$

$$y = -\frac{2}{3}x + b$$

$$6 = -\frac{2}{3}(-4) + b$$

$$6 = \frac{8}{3} + b$$

$$6 = \frac{10}{2} \rightarrow 5$$

$$\boxed{y = -\frac{2}{3}x + 5}$$

Composition is best explained with an example. Lets take two functions, f(x) and g(x), and plug g(x) INTO f(x).

We will also do the opposite and plug f(x) INTO g(x).

$f(x)=2x+3$	$f(x)=2x+3$
$g(x)=5x+2$	$g(x)=5x+2$
$(f\,o\,g)\,(x)=f(g(x))$	$(g\,o\,f)(x)=g(f(x))$
$(f\,o\,g)(x)=2(5x+2)+3$	$(g\,o\,f)(x)=5(2x+3)+2$
$(f\,o\,g)(x)=10x+4+3$	$(g\,o\,f)(x)=10x+15+2$
$\boxed{(f\,o\,g)(x)=10x+7}$	$\boxed{(f\,o\,g)(x)=10x+17}$

These compsitions can go both ways.

Notice the notation, in (f o g) the main function if "f" and the function we plug in is "g".

When it's (g o f) the main function is 'g' and the function we plug in is 'f'. We could also compose a function with itself.

This would look like this.

$$f(x) = 2x + 3$$

$$(f\,o\,f)(x) = 2(2x+3)+3$$

$$(f\,o\,f)(x) = 4x+6+3$$

$$\boxed{(f\,o\,f)(x) = 4x+9}$$

So as you can see, we just plugged the function into itself.

The Midpoint Formula

The midpoint formula calculates the...midpoint of the two points on the graph. The formula for this is..

$$M = \left(\frac{x_2 + x_1}{2}, \frac{y_2 + y_1}{2}\right)$$

Don't go nuts! There IS logic to this! As you can see your answer will be an ordered (x,y) pair.

And the reason for tha formula is this. Imagine you have your two x values and you want to take the average of them. You would add them up and divide by the number of values you have. THAT is the midpoint formula.

Think of it as the average value of x and the average value of y. Lets do an example:

Find the midpoint between (1,2) and (3,10).

This is the same ordered pair as before, but now we are asked to find its midpoint. Just plug in and evaluate.

$$M = \left(\frac{x_2 + x_1}{2}, \frac{y_2 + y_1}{2}\right)$$

$$M = \left(\frac{3+1}{2}, \frac{10+2}{2}\right)$$

$$M = \left(\frac{4}{2}, \frac{12}{2}\right)$$

$$\boxed{M = (2,6)}$$

Circles

The standard form of a circle is..

$$(x-h)^2 + (y-k)^2 = r^2$$

h and k represent the centre of the circle. (as an ordered pair) When there arent any numbers after x or y, then the circle is centred at the origin. The r stands for radius. Now dont get confused, if "r" said 25 the radius would be 5! Because thats the square root of 25. Now how did this equation come about? And what does it mean? Well lets imagine a circle.

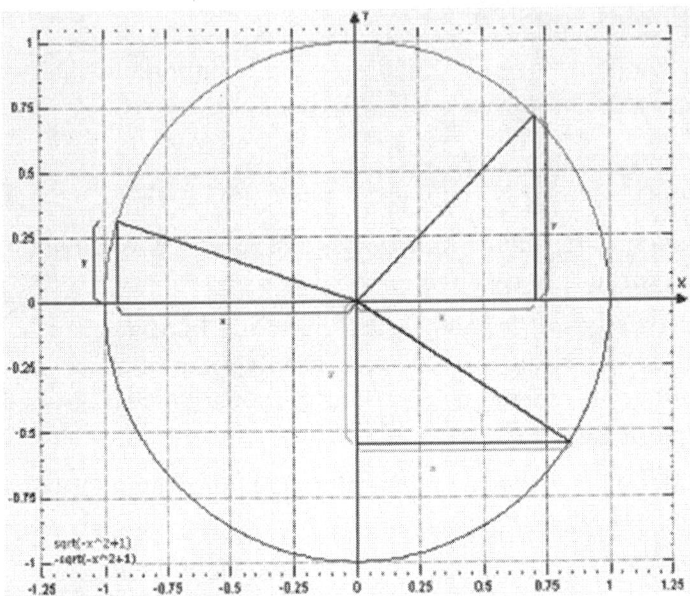

Think of x and y the same way we did when we talked about distance. The difference in the two horizontal distances, that would equal the base of the triangle on the coordinate plane. The same concept applies for y. And the hypotenuse of the triangle would bethe radius of the circle. So we can think of the equation for the circle as the pythagorean theorm that has all the solution points for the given radius. Lets look at some examples of the different types of circle equations you might see.

$$x^2 + y^2 = 25$$

This is a circle centred at the origin with a radius of 5.

$$(x - 2)^2 + (x + 4)^2 = 16$$

This is a circle with its centre at (2,–4) and a radius of 4.

Notice how the number next to the x tells you how many units in the x direction the circle moved, and the number next to the y tells you how many units in the y direction the circle moved.

But! It is in a direction you dont expect! For the same reasons as the transformations we talked about earlier. Because in the standard form there is a minus sign, so if you added a "–2" the two negatives would make it a plus. So just remember it's always opposite when it's in parenthesis!

Partial Fraction

In algebra, the partial fraction decomposition or partial fraction expansion is used to reduce the degree of *either* the numerator or the denominator of a rational function. The outcome of a full partial fraction expansion expresses that function as a sum of fractions, where:

- the denominator of each term is a power of an irreducible (not factorable) polynomial and
- the numerator is a polynomial of smaller degree than that irreducible polynomial.

In symbols, one can use *partial fraction expansion* to change a rational function in the form

$$\frac{f(x)}{g(x)}$$

where f and g are polynomials, into a function of the form

$$\sum \frac{a_n}{h_n(x)}$$

where h_n are polynomials that are factors of $g(x)$, and are in general of lower degree. The *full* decomposition pushes the reduction as far as it will go: in other words, the factorization of g is used as much as possible.

To decrease the degree of the numerator directly, the Euclidean algorithm can be used, but in fact if f already has lower degree than g this isn't helpful.

Thus the partial fraction decomposition may be seen as the inverse procedure of the more elementary operation of addition of fractions, that produces a single rational fraction with a numerator and denominator usually of high degree.

The main motivation to decompose a rational function into a sum of simpler fractions is that it makes simpler to perform linear operations on it. Therefore the problem of computing derivatives, antiderivatives, integrals, power expansions, Fourier series, residues, linear functional transformations, of rational functions can be reduced, via partial fraction decomposition, to making the computation on each single element used in the decomposition. See e.g. partial fractions in integration for an account of the use of the partial fractions in finding antiderivatives. Just which polynomials are irreducible depends on which field of scalars one adopts. Thus if one allows only real numbers, then irreducible polynomials are of degree either 1 or 2. If complex numbers are allowed, only 1st-degree polynomials can be irreducible. If one allows only rational numbers, or a finite field, then some higher-degree polynomials are irreducible.

Some Examples

The basic practical idea used for working with partial fractions is to work backwards. To separate a rational function as a sum, assume that there exist possible values of (unknown) parameters within a certain form. Then solve for these unknown by assuming equivalence to the initial function.

Simple Example

If one wants to decompose $x/(x + a)$, then one can follow these steps:

Write as

$$\frac{x}{x+a} = A + \frac{B}{x+a}$$

where : A, B and are constants

Multiply both sides by $x + a$

$$x = A(x+a)+B$$

$$x = A.x+A.a+B$$

means A and B must simultaneously solve:

$$\begin{cases} x = A.x \\ 0 = A.a + B \end{cases}$$

because the coefficients of both x and 1 have to agree, giving rise to the two equations.

Therefore

$$\begin{cases} A = 1 \\ B = -a. \end{cases}$$

Finally, the decomposed form is:

$$\frac{x}{x+a} = 1 + \frac{-a}{x+a}.$$

Distinct First-degree Factors in the Denominator

Suppose it is desired to decompose the rational function

$$\frac{x+3}{x^2-3x-40}$$

into partial fractions. The denominator factors as

$$(x-8)(x+5)$$

and so we seek scalars A and B such that

$$\frac{x+3}{x^2-3x-40}=\frac{x+3}{(x-8)(x+5)}=\frac{A}{x-8}+\frac{B}{x+5}.$$

One way of finding A and B begins by "clearing fractions", i.e., multiplying both sides by the common denominator $(x-8)(x+5)$. This yields

$$x+3=A(x+5)+B(x-8).$$

Collecting like terms gives

$$x+3=(A+B)x+(5a-8B)$$

Equating coefficients of like terms then yields:

$$A+B=1$$

$$5A-8B=3$$

The solution is $A = 11/13$, $B = 2/13$. Thus we have the partial fraction decomposition

$$\frac{x+3}{x^2-3x-40}=\frac{11/13}{8-x}+\frac{2/13}{x+5}=\frac{11}{13(x-8)}+\frac{2}{13(x+5)}$$

Alternatively, take the original equation

$$\frac{x+3}{(x-8)(x+5)}=\frac{A}{x-8}+\frac{B}{x+5}.$$

multiply by $(x-8)$ to get

$$\frac{x+3}{x+5}-A+\frac{B(x-8)}{x+5}.$$

evaluate at $x = 8$ to solve for A as

$$\frac{11}{13}=A.$$

multiply the original equation by $(x + 5)$ to get

$$\frac{x+3}{x-8} = \frac{A(x+5)}{x-8} + B.$$

evaluate at $x = -5$ to solve for B as

An Irreducible Quadratic Factor in the Denominator

In order to decompose

$$\frac{10x^2+12x+20}{x^3-8}$$

into partial fractions, first observe that

$$x^3-8=(x-2)(x^2+2x+4).$$

The fact that $x^2 + 2x + 4$ cannot be factored using real numbers can be seen by observing that the discriminant $2^2 - 4(1)(4)$ is negative. Thus we seek scalars A, B, C such that

$$\frac{10x^2+12x+20}{x^3-8} = \frac{10x^2+12x+20}{(x-2)(x^2+2x+4)} = \frac{A}{x-2} + \frac{Bx+C}{x^2+2x+4}.$$

When we clear fractions, we get

$$10x^2+12x+20 = A(x^2+2x+4)+(Bx+C)(x-2).$$

We could proceed as in the previous example, getting three linear equations in three variables A, B, and C. However, since solving such systems becomes onerous as the number of variables grows, we try a different method. Substitution of 2 for x in the identity above makes the entire second term vanish, and we get

$$10.2^2+12.2+20 = A(2^2+2.2+4),$$

i.e., $84 = 12A$, so $A = 7$, and we have

$$10x^2+12x+20 = 7(x^2+2x+4)+(Bx+C)(x-2).$$

Next, substitution of 0 for x yields

$$20 = 7(4)+C(-2),$$

and so $C = 4$. We now have

$$10x^2+12x+20 = 7(x^2+2x+4)+(Bx+4)(x-2).$$

Substitution of 1 for x yields

$$10+12+20=7(1+2+4)+(B+4)(1-2),$$

and so B = 3. Our partial fraction decomposition is therefore:

$$\frac{10x^2+12x+20}{x^3-8}=\frac{7}{x-2}+\frac{3x+4}{x^2+2x+4}.$$

A Repeated First-degree Factor in the Denominator

Consider the rational function

$$\frac{10x^2-63x+29}{x^3-11x^x+40x-48}.$$

The denominator factors thus:

$$x^3-11x^2+40x-48=(x-3)(x-4)^2.$$

The multiplicity of the first-degree factor (x ” 4) is more than 1. In such cases, the partial fraction decomposition takes the following form:

$$\frac{10x^2-63x+29}{x^3-11x^2+40x-48}=\frac{10x^2-63x+29}{(x-3)(x-4)^2}$$

$$=\frac{A}{x-3}+\frac{B}{x-4}+\frac{C}{(x-4)^2}.$$

By solving, A = –70, B = 80, C = –63. Therefore, the solution becomes

$$\frac{10x^2-63x+29}{x^3-11x^2+40x-48}=\frac{-70}{x-3}+\frac{80}{x-4}+\frac{-63}{(x-4)^2}.$$

Repeated Factors in the Denominator Generally

For rational functions of the form

$$\frac{p(x)}{(x+2)(x+3)^5}$$

(where the $p(x)$ may be any polynomial of sufficiently small degree) the partial fraction decomposition looks like this:

$$\frac{A}{x+2}+\frac{B}{x+3}+\frac{C}{(x+3)^2}+\frac{D}{(x+3)^3}+\frac{E}{(x+3)^4}+\frac{F}{(x+3)^5}.$$

The general pattern may be quickly guessed. For rational functions of the form

$$\frac{p(x)}{(x+2)(x^2+1)^5}$$

with the irreducible quadratic factor $x^2 + 1$ in the denominator (where again, the $p(x)$ may be any polynomial of sufficiently small degree), the partial fraction decomposition looks like this:

$$\frac{A}{x+2}+\frac{Bx+C}{x^2+1}+\frac{Dx+E}{(x^2+1)^2}+\frac{Fx+G}{(x^2+1)^3}+\frac{Hx+I}{(x^2+1)^4}+\frac{Jx+K}{(x^2+1)^5}.$$

and a similar pattern holds for any other irreducible quadratic factor.

High-degree Polynomials in the Numerator

When you need to apply the partial fraction decomposition to a polynomial division like

$$\frac{a_n x^n + \ldots + a_1 x + a_0}{b_m x^m + \ldots + b_1 x + b_0}$$

where $n \geq m$, you just need to apply the polynomial long division procedure first, and apply the partial fraction decomposition to the remainder.

Use in Deriving the Logistic General Equation

In many beginning calculus courses, partial fractions are introduced as a way to derive the general equation for a logistic function. Logistic functions model a population which grows until it reaches a limit.

The rate of change for the function is proportional (constant k) to both the population reached (P) and the fraction of the total carrying capacity (M) remaining. Thus:

$$\frac{dP}{dt} = kP\left(\frac{M-P}{M}\right)$$

$$\int \frac{1}{P(M-P)}\,dP = \int \frac{k}{M}\,dt$$

$$\frac{1}{P(M-P)} = \frac{A}{P} + \frac{B}{(M-P)}$$

$$1 = A(M-P) + BP$$

$$1 = AM - AP + BP$$

$$A = B, A = \frac{1}{M}, B = \frac{1}{M}$$

$$\int \frac{1}{MP} + \frac{1}{M(M-P)} dP = \int \frac{k}{M} dt$$

$$\int \frac{1}{P} + \frac{1}{(M-P)} dP = \int k\, dt$$

$$\ln \frac{P}{(P-M)} = kt + C$$

$$\frac{P}{P-M} = e^{kt+C}$$

$$\frac{M-P}{P} = e^{-kt-C}$$

$$\frac{M}{P} - 1 = e^{-kt-C}$$

$$P = \frac{M}{e^{-kt-C} + 1}$$

$$P = \frac{M}{Ae^{-kt} + 1}$$

Basic Principles

The basic principles involved are quite simple; it is the algorithmic aspects that require attention in particular cases. On the other hand, the existence of a decomposition of a certain kind is an assumption in practical cases, and the principles should explain which assumptions are justified.

Assume a rational function $R(x)$ in one indeterminate x has a denominator that factors as

$$P(x)Q(x)$$

over a field K (we can take this to be real numbers, or complex numbers). If P and Q have no common factor, then R may be written as

$$A/P + B/Q$$

for some polynomials $A(x)$ and $B(x)$ over K. The *existence* of such a decomposition is a consequence of the fact that the polynomial ring over K is a principal ideal domain, so that

$$CP + DQ = 1$$

for some polynomials $C(x)$ and $D(x)$ (see Bézout's identity).

Using this idea inductively we can write $R(x)$ as a sum with denominators powers of irreducible polynomials. To take this further, if required, write

$$G(x)/F(x)^n$$

as a sum with denominators powers of F and numerators of degree less than F, plus a possible extra polynomial. This can be done by the Euclidean algorithm, polynomial case.

Therefore when K is the complex numbers and we can assume F has degree 1 (by the fundamental theorem of algebra) the numerators will be constant.

When K is the real numbers we can have the case of

$$\text{degree } F = 2,$$

and a quotient of a linear polynomial by a power of a quadratic will occur. This therefore is a case that requires discussion, in the systematic theory of integration (for example in computer algebra).

The Role of the Taylor Polynomial

The partial fraction decomposition of a rational function can be related to Taylor's theorem as follows. Let

$$P(x), Q(x), A_1(x), \ldots, A_r(x)$$

be real or complex polynomials; assume that

$$Q = \prod_{j=1}^{r} (x - \lambda_j)^{\nu_j},$$

that

$$\deg(P) < \deg(Q) = \sum_{j=1}^{r} \nu_j,$$

and that

$$\deg A_j < \nu_j \text{ for } j = 1, \ldots, r.$$

Define also

$$Q_i = \prod_{j \neq i} (x - \lambda_j)^{\nu_j} = \frac{Q}{(x - \lambda_i)^{\nu_i}} \text{ for } i = 1, \ldots, r.$$

Then we have

$$\frac{P}{Q}=\sum_{j=1}^{r}\frac{A_j}{(x-\lambda_j)^{\nu_j}}$$

if, and only if, for each i the polynomial $A_i(x)$ is the Taylor polynomial of $\frac{P}{Q_i}$ of order ν_i-1 at the point λ_i:

$$A_i(x):=\sum_{k=0}^{\nu_i-1}\frac{1}{k!}\left(\frac{P}{Q_i}\right)^{(k)}(\lambda_i)\,(x-\lambda_i)^k.$$

Taylor's theorem (in the real or complex case) then provides a proof of the existence and uniqueness of the partial fraction decomposition, and a characterization of the coefficients.

Sketch of the proof: The above partial fraction decomposition implies, for each $1\le i\le r$, a polynomial expansion

$$\frac{P}{Q_i}=A_i+O((x-\lambda_i)^{\nu_i})$$

as $x\to\lambda_i$;

so A_i is the Taylor polynomial of $\frac{P}{Q_i}$, because of the unicity of the polynomial expansion of order ν_i-1, and by assumption $\deg A_i<\nu_i$.

Conversely, if the A_i are the Taylor polynomials, the above expansions at each hold λ_i, therefore we also have

$$P-Q_iA_i=O((x-\lambda_i)^{\nu_i})\qquad \text{as } x\to\lambda_i,,$$

which implies that the polynomial $P-Q_iA_i$ is divisible by $(x-\lambda_i)^{\nu_i}$.

For $j\neq i$ also Q_jA_j is divisible by $(x-\lambda_i)^{\nu_i}$, so we have in turn that $P-\sum_{j=1}^{r}Q_jA_j$ is divisible by Q. Since $\deg\left(P-\sum_{j=1}^{r}Q_jA_j\right)<\deg(Q)$ we then have $P-\sum_{j=1}^{r}Q_jA_j=0$, and we find the partial fraction decomposition dividing by Q.

Algorithms

Lagrange Interpolation

Partial fraction decomposition can be derived using Lagrange interpolation.

Example

As an introductory example we take the rational function

$$\frac{x}{x^2-1}.$$

By the difference of two squares identity, this can also be written as

$$\frac{x}{(x+1)(x-1)},$$

which can be transformed further. Consider an identity

$$\frac{A}{x+1}+\frac{B}{x-1}=\frac{x}{x^2-1},$$

where A and B are constants. In more explicit form, we have the relation of the numerators,

$$A(x-1)+B(x+1)=x.$$

We know that the constants on one side of an expression must equal those on the other side.

On the left hand side, the constants are "A and B, and on the right, the constant is simply 0. So, comparing constants on both sides of the expression, we can see that

$$B-A=0, \text{ i.e. } A=B.$$

Now, in the same way, we know that the number of x terms on the left must equal the number of x's on the right. Therefore, looking at x terms on both sides,

$$Ax+Bx=x,$$

therefore

$$A+B=1,$$

and so, given that $A = B$, we can say that

$$A+A=1$$

$$2A=1$$

$$A=B=0.5.$$

Finally we find:

$$\frac{x}{x^2-1}=\frac{1}{2}.\frac{1}{x+1}+\frac{1}{2}.\frac{1}{x-1}$$

or

$$\frac{x}{x^2-1}=\frac{1}{2(x+1)}+\frac{1}{2(x-1)}$$

which holds true for all $x \neq \pm 1$.

Derivation

The preceding example can be generalized to the following situation:

Assume that $Q(x)$ is a monic polynomial of some degree n which over the underlying field K decomposes into linear factors

$$Q(x)=\prod_{i=1}^{n}(x-x_i),$$

where all x_i are pairwise different. In other words Q has simple roots (over K). If $P(x)$ is any polynomial of degree $\leq n-1$ then according to the Lagrange interpolation formula $P(x)$ can be uniquely written as a sum (the Lagrange form representation)

$$P(x)=\sum_{j=1}^{n}P(x_j)L_j(x;x_j),$$

where $L_j(x;x_j)$ is the Lagrange polynomial

$$L_j\left(x;x_j\right)=\prod_{k\leq n,k\neq j}\frac{\left(x-x_k\right)}{\left(x_j-x_k\right)}.$$

Dividing the Lagrange representation on the right side termwise by the polynomial $Q(x)$ in its factored form one obtains

$$\frac{P(x)}{Q(x)}=\sum_{j=1}^{n}\frac{P\left(x_j\right)}{\prod k\leq n,k\neq j\left(x_j-x_k\right)}\cdot\frac{1}{x-x_j}.$$

This is the partial fraction decomposition

$$\frac{P(x)}{Q(x)}=\sum_{j=1}^{n}\frac{P\left(x_j\right)}{Q\left(x_j\right)}\cdot\frac{1}{x-x_j}$$

of the rational function $R(x)=P(x)/Q(x)$ with the derivative of Q given by:

$$Q'\left(x_j\right)=\sum_{I\leq n}\prod_{k\leq n,k\neq I}\left(x_j-x_k\right)=\prod_{k\leq n,k\neq j}\left(x_j-x_k\right).$$

The first example can be obtained as the special case

$$Q(x) = (x-1).(x+1), P(x) = x.$$

Note the close relationship to divided differences.

ParAvartya Sutra

Separation of a fractional algebraic expression into partial fractions is the reverse of the process of combining fractions by converting each fraction to the lowest common denominator (LCM) and adding the numerators. This separation is accomplished by a mental, one-line Vedic formula called the *Paravartya Sutra*. Case one has fractional expressions where factors in the denominator are unique.

Case two has fractional expressions where some factors may repeat as powers of a binomial. In integral calculus we would want to write a fractional algebraic expression as the sum of its partial fractions in order to take the integral of each simple fraction separately.

Once the original denominator, D_0, has been factored we set up a fraction for each factor in the denominator. We may use a subscripted D to represent the denominator of the respective partial fractions which are the factors in D_0. Letters A, B, C, D, E, and so on will represent the numerators of the respective partial fractions.

We calculate each respective numerator by (1) calculating the *Parâvartya* value of the denominator (which is the value of the variable making that binomial factor equal to zero) and (2) then substituting this value into the original expression but ignoring that factor in the denominator.

Each *Parâvartya* value for the variable is the value which would give an undefined value to the expression since we do not divide by zero.

General Formula:

$$\frac{\ell x^2 + mx + n}{(x-a)(x-b)(x-c)} = \frac{A}{(x-a)} + \frac{B}{(x-b)} + \frac{C}{(x-c)}$$

Here, a, b, c,, m, and n are given integer values.

Where $x = a$ and

$$A = \frac{\ell a^2 + ma + n}{(a-b)(a-c)};$$

and where $x = b$ and

$$B = \frac{\ell b^2 + mb + n}{(b-c)(b-a)};$$

and where $x = c$ and

$$C = \frac{\ell c^2 + mc + n}{(c-a)(c-b)};$$

Case One

Factorize the expression in the denominator. Set up a partial fraction for each factor in the denominator. Apply the *Parâvartya Sûtra* to solve for the new numerator of each partial fraction.

Example

$$\frac{3x^2 + 12x + 11}{(x+1)(x+2)(x+3)} = \frac{A}{x+1} + \frac{B}{x+2} + \frac{C}{x+3}$$

Set up a partial fraction for each factor in the denominator. With this framework we apply the Sûtra to solve for A, B, and C by mental math.

1. D_1 is $x + 1$; set it equal to zero. This gives the *Parâvartya* value for A when $x = -1$.
2. Next, substitute this value of x into the fractional expression, but without D_1.
3. Put this value down as the value of A.

Proceed similarly for B and C.

D_2 is $x + 2$; For *Parâvartya* B use $x = -2$.

D_3 is $x + 3$; For *Parâvartya* C use $x = -3$.

Thus, to solve for A, use $x = -1$ in the expression but without D_1:

$$\frac{3x^2 + 12x + 11}{(x+2)(x+3)} = \frac{3 - 12 + 11}{(1)(2)} = \frac{2}{2} = 1 = A.$$

Thus, to solve for B, use $x = -2$ in the expression but without D_2:

$$\frac{3x^2 + 12x + 11}{(x+1)(x+3)} = \frac{12 - 24 + 11}{(-1)(1)} = \frac{-1}{(-1)} = +1 = B.$$

Thus, to solve for C, use $x = -3$ in the expression but without D_3:

$$\frac{3x^2 + 12x + 11}{(x+1)(x+2)} = \frac{27 - 36 + 11}{(-2)(-1)} = \frac{2}{(+1)} = +1 = C.$$

Thus,

$$\frac{3x^2+12x+11}{(x+1)(x+2)(x+3)}=\frac{1}{x+1}+\frac{1}{x+2}+\frac{1}{x+3}$$

Case Two

When factors of the denominator include powers of one expression we

(1) Set up a partial fraction for each unique factor and each lower power of D;

(2) We set up an equation showing the relation of the numerators if all were converted to the LCD.

From the equation of numerators we solve for each numerator, A, B, C, D, and so on. This equation of the numerators is an absolute identity, true for all values of x. So, we may select any value of x and solve for the numerator.

Example

$$\frac{3x+5}{(1-2x)^2}=\frac{A}{(1-2x)^2}+\frac{B}{(1-2x)}$$

Here, we set up a partial fraction for each descending power of the denominator. Then we solve for the numerators, A and B. As the *Parâvartya* value for A and B will be the same, $x = ½$, we need an additional relation in order to solve for both.

To write the relation of numerators the second fraction needs another factor of (1–2x) to convert it to the LCD, giving us $3x + 5 = A + B(1 - 2x)$.

To solve for A: Set the denominator of the first fraction to zero, $1 - 2x = 0$. Solving for x gives the *Parâvartya* value for A, when $x =$ ½. When we substitute this value, $x =$ ½, into the relation of numerators we have $3(1/2) + 5 = A + B(0)$. Solving for A gives us $A = 3/2 + 5 = 13/2$. Hence, numerator A equals six and one-half.

To solve for B: Since the equation of the numerators, here,

$$3x + 5 = A + B(1 - 2x),$$

is true for all values of x, pick a value for x and use it to solve for B. As we have solved for the value of A above, $A = 13/2$, we may use that value to solve for B.

We may pick $x = 0$, use $A = 13/2$, and then solve for B.

$3x + 5 = A + B(1 \text{ " } 2x)$

$0 + 5 = 13/2 + B(1 + 0)$

$10/2 = 13/2 + B$

$-3/2 = B$

We may pick $x = 1$. Then solve for B:

$3x + 5 = A + B(1 - 2x)$

$3 + 5 = 13/2 + B(1 - 2)$

$8 = 13/2 + B(-1)$

$16/2 = 13/2 - B$

$B = -3/2.$

We may pick $x = -1$. Solve for B:

$3x + 5 = A + B(1 - 2x)$

$-3 + 5 = 13/2 + B(1 + 2)$

$4/2 = 13/2 + 3B$

$-9/2 = 3B$

$-3/2 = B$

Hence,

$$\frac{3x+5}{(1-2x)^2} = \frac{13/2}{(1-2x)^2} + \frac{-3/2}{(1-2x)},$$

or

$$\frac{3x+5}{(1-2x)^2} = \frac{13}{2(1-2x)^2} - \frac{3}{2(1-2x)}.$$

Technique Three

A third technique is an analysis of the expanded relation of the numerators. Just match up the x-terms of each degree. Then one can see the coefficients of the matching terms and solve for the missing numerator.

Example

$$\frac{3x+5}{(1-2x)^2} = \frac{A}{(1-2x)^2} + \frac{B}{(1-2x)}$$

Converting each fraction to the LCD we have the relation in the numerators: $3x + 5 = A + B(1 - 2x)$. As A and B are constants, we can expand and match up the constant terms and the x–terms. The values of A and B will then be apparent.

$$3x + 5 = A + B(1 - 2x)$$

$$3x + 5 = A + B - 2Bx$$

Hence, the constant terms are set equal and the x-terms are set equal:

$$5 = A + B \text{ and } 3x = -2Bx$$

Therefore, by setting the coefficients equal, we may solve for B:

$$3 = -2B$$

$$-3/2 = B$$

And then solve for A:

$$5 = A - 3/2$$

$$5 + 3/2 = A$$

$$13/2 = A$$

Hence,

$$\frac{3x+5}{(1-2x)^2} = \frac{13/2}{(1-2x)^2} + \frac{-3/2}{(1-2x)}.$$

Fractions of Integers

The idea of partial fractions can be generalized to other rings, say the ring of integers where prime numbers take the role of irreducible denominators. E.g., it is:

$$\frac{1}{18} = \frac{1}{2} - \frac{1}{3} - \frac{1}{3^2}.$$

Partial Fraction Decomposition

Partial fraction decomposition is a theorem in algebra which states that a rational function over a field can be decomposed into a polynomial plus a sum of proper fractions, each of which has the form $p(x)/q(x)^n$, where the degree of $p(x)$ is smaller than the degree of $q(x)$ and $q(x)$ is irreducible.

Statement of Theorem

Let f and g be nonzero polynomials. Write g as a product of powers of distinct irreducible polynomials:

$$g = \prod_{i=1}^{k} p_i^{ni}.$$

There are (unique) polynomials b and a_{ij} with deg a_{ij} < deg p_i such that

$$\frac{f}{g}=b+\sum_{i=1}^{k}\sum_{j=1}^{ni}\frac{a_{ij}}{p_i^j}.$$

If deg f < deg g, then $b = 0$.

Outline of Proof

Lemma 1

Let f, g and h be nonzero polynomials with f and g coprime. There are polynomials a and b such that $\frac{h}{fg}=\frac{a}{f}+\frac{b}{g}$.

Lemma 2

Let f and g be nonzero polynomials and let n be a positive integer. There exist polynomials b and a_i with deg a_i < deg g such that

$$\frac{f}{g^n}=b+\sum_{j=1}^{n}\frac{a_j}{g^j}.$$

Over the Reals

In mathematics, partial fractions are used in real-variable integral calculus to find real-valued antiderivatives of rational functions. Partial fraction decomposition of real rational functions is also used to find their Inverse Laplace transforms.

This article describes the general method for obtaining the partial fraction decomposition of any *real* rational function.

The proof of the existence of the partial fraction decomposition over an arbitrary field is not given here. For a sketch of the general proof, see partial fraction.

For applications of partial fraction decomposition over the reals, see

- Partial fractions in integration
- Partial fractions in Laplace transforms

First we state the general result, and then we offer several examples to illustrate the result and show how the method is used in practice.

General Result

Let $f(x)$ be any rational function over the real numbers. In other words, suppose there exist real polynomials $p(x)$ and $q(x) \neq 0$, such that

$$f(x) = \frac{p(x)}{q(x)}$$

By removing the leading coefficient of $q(x)$, we may assume without loss of generality that $q(x)$ is monic. By the fundamental theorem of algebra, we can write

$$q(x) = (x - a_1)^{j1} \ldots (x - a_m)^{jm} (x^2 + b_1 x + c_1)^{k1} \ldots (x^2 + b_n x + c_n)^{k_n}$$

where $a_1, \ldots, a_m, b_1, \ldots, b_n, c_1, \ldots, c_n$ are real numbers with $b_i^2 - 4c_i < 0$, and $j_1, \ldots, j_m, k_1, \ldots, k_n$ are positive integers. The terms $(x - a_i)$ are the *linear factors* of $q(x)$ which correspond to real roots of $q(x)$, and the terms $(x_i^2 + b_i x + c_i)$ are the *irreducible quadratic factors* of $q(x)$ which correspond to pairs of complex conjugate roots of $q(x)$.

Then the partial fraction decomposition of $f(x)$ is the following.

$$f(x) = \frac{p(x)}{q(x)} = P(x) + \sum_{i=1}^{m} \sum_{r=1}^{j_i} \frac{A_{ir}}{(x - a_i)r} + \sum_{i=1}^{n} \sum_{r=1}^{k_i} \frac{B_{ir} x + C_{ir}}{(x^2 + b_i x + c_i)^r}$$

Here, $P(x)$ is a (possibly zero) polynomial, and the A_{ir}, B_{ir}, and C_{ir} are real constants. There are a number of ways the constants can be found.

The most straightforward method is to multiply through by the common denominator $q(x)$. We then obtain an equation of polynomials whose left-hand side is simply $p(x)$ and whose right-hand side has coefficients which are linear expressions of the constants A_{ir}, B_{ir}, and C_{ir}. Since two polynomials are equal if and only if their corresponding coefficients are equal, we can equate the coefficients of like terms.

In this way, a system of linear equations is obtained which *always* has a unique solution. This solution can be found using any of the standard methods of linear algebra.

However, this is often not the best way to go when computing by pencil and paper, and there are other ways ("tricks") to obtain the constants.

The Algorithm

The partial fraction decomposition of a real rational function can be obtained by the following procedure:

1. Remove the leading coefficient of the denominator $q(x)$ if necessary.
2. Reduce the rational function as much as possible; i.e. cancel any obvious common factors of the numerator and denominator.
3. Make sure the degree of the numerator $p(x)$ is *strictly* smaller than the degree of the denominator $q(x)$.

   If it is not, then use the division algorithm for polynomials and long-divide the denominator into the numerator. You will get a (nonzero) polynomial quotient and a remainder whose degree is now strictly smaller than the degree of the denominator. Take this remainder as your new numerator and continue with the following steps. (Don't forget the polynomial quotient when you're done!)
4. Factor the denominator $q(x)$ into linear and irreducible quadratic factors. This is not a straightforward task; however, you have some tools to use. If you find any real root a of $q(x)$, then you know that $(x–a)$ is a factor of $q(x)$.

   Also, if you differentiate $q(x)$ and use the Euclidean algorithm to find the greatest common divisor of $q(x)$ and $q(x)$, then any real root a of the greatest common divisor gives you a factor $(x–a)$. You can look for ways to complete the square in different powers of x. Most importantly, you can use the rational root test to determine all possible rational roots of $q(x)$.
5. Express the partial fraction decomposition of $f(x)$ as the sum of double sums above.
6. Find the constants A_{ir}, B_{ir}, and C_{ir} by whatever methods possible. Once you have found all the constants, you now have the partial fraction decomposition. (Don't forget the polynomial quotient!)

Some Examples

Example 1:

$$f(x) = \frac{1}{x^2 + 2x - 3}$$

Here, the denominator splits into two distinct linear factors:

$$q(x) = x^2 + 2x - 3 = (x + 3)(x - 1)$$

so we have the partial fraction decomposition

$$f(x)=\frac{1}{x^2+2x-3}=\frac{A}{x+3}+\frac{B}{x-1}$$

Multiplying through by $x^2 + 2x - 3$, we have the polynomial identity

$$1 = A(x - 1) + B(x + 3)$$

Substituting $x = -3$ and $x = 1$ into this equation gives

$A = -1/4$ and $B = 1/4$, so that

$$f(x)=\frac{1}{x^2+2x-3}=\frac{A}{x+3}\frac{1}{4}\left(\frac{-1}{x+3}+\frac{1}{x-1}\right)$$

Example 2

$$f(x)=\frac{x^3+16}{x^3-4x^2+8x}$$

After long-division, we have

$$f(x)=1+\frac{4x^2-8x+16}{x^3-4x^2+8x}=1+\frac{4x^2-8x+16}{x(x^2-4x+8)}$$

Since $(-4)^2 - 4(8) = -16 < 0$, $x^2 - 4x + 8$ is irreducible, and so

$$\frac{4x^2-8x+16}{x(x^2-4x+8)}=\frac{A}{x}+\frac{Bx+C}{x^2-4x+8}$$

Multiplying through by $x^3 - 4x^2 + 8x$, we have the polynomial identity

$$4x^2 - 8x + 16 = A(x^2 - 4x + 8) + (Bx + C)x$$

Taking $x = 0$, we see that $16 = 8A$, so $A = 2$. Comparing the x^2 coefficients, we see that $4 = A + B = 2 + B$, so $B = 2$. Comparing linear coefficients, we see that $-8 = -4A + C = -8 + C$, so $C = 0$. Altogether,

$$f(x)=1+2\left(\frac{1}{x}+\frac{x}{x^2-4x+8}\right)$$

The following example illustrates almost all the "tricks" one would need to use short of consulting a computer algebra system.

Example 3

$$f(x)=\frac{x^9-2x^6+2x^5-7x^4+13x^3-11x^2+12x-4}{x^7-3x^6+5x^5-7x^4+7x^3-5x^2+3x-1}$$

After long-division and factoring, we have

$$f(x) = x^2 + 3x + 4x + \frac{x^5 - 2x^4 + 5x^3 - 5x^2 + 6x - 1}{(x-1)^3(x^2+1)^2}.$$

The partial fraction decomposition takes the form

$$\frac{x^5 - 2x^4 + 5x^3 - 5x^2 + 6x - 1}{(x-1)^3(x^2+1)^2} = \frac{A}{x-1} + \frac{B}{(x-1)^2} + \frac{C}{(x-1)^3} + \frac{Dx+E}{x^2+1} + \frac{Fx+G}{(x^2+1)^2}$$

Multiplying through by $(x - 1)^3(x^2 + 1)^2$ we have the polynomial identity

$$x^5 - 2x^4 + 5x^3 - 5x^2 + 6x - 1$$
$$= A(x - 1)^2(x^2 + 1)^2 + B(x - 1)(x^2 + 1)^2 + C(x^2 + 1)^2$$
$$+ (Dx + E)(x - 1)^3(x^2 + 1) + (Fx + G)(x - 1)^3$$

Taking $x = 1$ gives $4 = 4C$, so $C = 1$.

Similarly, taking $x = i$ gives $2 + 2i = (Fi + G)(2 + 2i)$, so $Fi + G = 1$, so $F = 0$ and $G = 1$ by equating real and imaginary parts.

We now have the identity:

$$x^5 - 2x^4 + 5x^3 - 5x^2 + 6x - 1$$
$$= A(x - 1)^2(x^2 + 1)^2 + B(x - 1)(x^2 + 1)^2 + (x^2 + 1)^2$$
$$+ (Dx + E)(x - 1)^3(x^2 + 1) + (x - 1)^3$$

Taking constant terms gives $E = A - B + 1$, taking leading coefficients gives $A = -D$, and taking x-coefficients gives $B = 3 - D - 3E$.

Putting all of this together,

$$E = A-B+1=-D-(3-D-3E)+1=3E-2,$$

so

$$E = 1 \text{ and}$$
$$A = B = -D.$$

Now,

$$x^5 - 2x^4 + 5x^3 - 5x^2 + 6x - 1$$
$$= A(x - 1)^2(x^2 + 1)^2 + A(x - 1)(x^2 + 1)^2$$
$$+ (x^2 + 1)^2 + (- Ax + 1)(x - 1)^3(x^2 + 1) + (x - 1)^3$$

Taking $x = -1$ gives $-20 = -8A - 20$, so $A = B = D = 0$. The partial fraction decomposition of $f(x)$ is thus:

$$f(x) = x^2 + 3x + 4 + \frac{1}{(x-1)^3} + \frac{1}{x^2+1} + \frac{1}{(x^2+1)^2}$$

General Methods for Finding Coefficients

We present in the sequel a list of methods for finding the coefficients occurring in the r.h.s. sums of the above equation for $f(x)=p(x)/q(x)=\ldots$, to which we shall refer to by (PFD).

For some of the methods it would be useful if p and/or q were already written in factored form.

Simple Poles of Multiplicity 1

If a_i is a simple pole of f, i.e. $j_i = 1$, then one multiplies (PFD) by $(x ” a_i)$ and takes the limit $x \to a_i$. On the r.h.s. only A_{i1} "survives", which is therefore given by

$$A_{i1} = \lim_{x \to a} i \ (x \text{ &minuns; } a_i)\ f(x) = p(a_i)/q_i(a_i)$$

where q_i is q divided (simplified) by the factor $(x ” a_i)$. It holds $q_i(a_i) = q'(a_i)$.

Leading Term of a Pole

Using the same idea, we determine the coefficients A_{ij_i} (multiplying in the highest negative power), by just replacing $(x ” a_i)$ by its j_ith power.

Leading Term of a Quadratic Factor

Still using the same idea, but going to the complex root of $x^2 + bx + c$, one gets in the same way coefficients B_{ik_i}, C_{ik_i} (from real and imaginary part of the previous equation for A which is now an equation for $Bx + C$).

Other Coefficients A_{ik}

All coefficients relative to one pole can be obtained "in one step" by making a change of variables $t = x - a_i$ (i.e. replacing x by $t + a_i$) and then (long) dividing the "new" p by the "new" q, with the "pure" power $t^j{}_i$ removed.

Method of Derivatives

Instead of making the change of variables and using long division, one can obtain a recursive formula involving derivatives of p and q_i at $x = a_i$.

General Methods

Besides the above methods to obtain a particular coefficient, there are several general methods to get one or more equations involving one or several of the coefficients, which should then allow to determine the remaining unknown coefficients:

- method of particular values: this just consists in putting x equal to particular fixed values (0, ± 1, ± 2...), and thus obtaining numerical equations for the coefficients.
- method of limits: multiply both sides of (PFD), after subtracting known terms (especially the polynomial), by the highest possible power of x such that lim(x→") still is finite, and take this limit
- method of identification: the "final rescue", put all of the r.h.s. on the common denominator $q(x)$ and identify coefficients of alike powers in the numerator on the r.h.s. and $p(x)$ on the l.h.s.

6

Set Theory

Set theory is the branch of mathematics that studies sets, which are collections of objects. Although any type of object can be collected into a set, set theory is applied most often to objects that are relevant to mathematics.

The modern study of set theory was initiated by Cantor and Dedekind in the 1870s. After the discovery of paradoxes in informal set theory, numerous axiom systems were proposed in the early twentieth century, of which the Zermelo–Fraenkel axioms, with the axiom of choice, are the best-known.

The language of set theory is used in the definitions of nearly all mathematical objects, such as functions, and concepts of set theory are integrated throughout the mathematics curriculum. Elementary facts about sets and set membership can be introduced in primary school, along with Venn diagrams, to study collections of commonplace physical objects.

Elementary operations such as set union and intersection can be studied in this context. More advanced concepts such as cardinality are a standard part of the undergraduate mathematics curriculum.

Set theory, formalized using first-order logic, is the most common foundational system for mathematics. Beyond its use as a foundational system, set theory is a branch of mathematics in its own right, with an active research community.

Contemporary research into set theory includes a diverse collection of topics, ranging from the structure of the real number line to the study of the consistency of large cardinals.

History

Mathematical topics typically emerge and evolve through interactions among many researchers. Set theory, however, was founded by a single paper in 1874 by Georg Cantor: "On a Characteristic Property of All Real Algebraic Numbers".

Beginning with the work of Zeno around 450 BC, mathematicians had been struggling with the concept of infinity. Especially notable is the work of Bernard Bolzano in the first half of the 19th century.

The modern understanding of infinity began 1867-71, with Georg Cantor's work on number theory. An 1872 meeting between Cantor and Richard Dedekind influenced Cantor's thinking and culminated in Cantor's 1874 paper.

Cantor's work initially polarized the mathematicians of his day. While Weierstrass and Dedekind supported Cantor, Kronecker, now seen as a founder of mathematical constructivism, did not.

Cantorian set theory eventually became widespread, due to the utility of Cantorian concepts, such as one-to-one correspondence among sets, his proof that there are more real numbers than integers, and the "infinity of infinities" ("Cantor's paradise") the power set operation gives rise to.

The next wave of excitement in set theory came around 1900, when it was discovered that Cantorian set theory gave rise to several contradictions, called antinomies or paradoxes. Russell and Zermelo independently found the simplest and best known paradox, now called Russell's paradox and involving "the set of all sets that are not members of themselves."

This leads to a contradiction, since it must be a member of itself and not a member of itself. In 1899 Cantor had himself posed the question: "what is the cardinal number of the set of all sets?" and obtained a related paradox.

The momentum of set theory was such that debate on the paradoxes did not lead to its abandonment. The work of Zermelo in 1908 and Fraenkel in 1922 resulted in the canonical axiomatic set theory ZFC, which is thought to be free of paradoxes.

The work of analysts such as Lebesgue demonstrated the great mathematical utility of set theory. Axiomatic set theory has become woven into the very fabric of mathematics as we know it today.

Basic Concepts

Set theory begins with a fundamental binary relation between an object o and a set A. If o is a member (or element) of A, we write $o \in A$. Since sets are objects, the membership relation can relate sets as well.

A derived binary relation between two sets is the subset relation, also called set inclusion. If all the members of set A are also members of set B, then A is a subset of B, denoted $A \subseteq B$. For example, {1,2} is a subset of {1,2,3}, but {1,4} is not. From this definition, it is clear that a set is a subset of itself; in cases where one wishes to avoid this, the term proper subset is defined to exclude this possibility.

Just as arithmetic features binary operations on numbers, set theory features binary operations on sets. The:

- Union of the sets A and B, denoted $A \cup B$, is the set whose members are members of at least one of A or B. The union of {1, 2, 3} and {2, 3, 4} is the set {1, 2, 3, 4}.
- Intersection of the sets A and B, denoted $A \cup B$, is the set whose members are members of both A and B. The intersection of {1, 2, 3} and {2, 3, 4} is the set {2, 3}.
- Complement of set A relative to set U, denoted A^c, is the set of all members of U that are not members of A. This terminology is most commonly employed when U is a universal set, as in the study of Venn diagrams. This operation is also called the set difference of U and A, denoted $U \setminus A$. The complement of {1,2,3} relative to {2,3,4} is {4}, while, conversely, the complement of {2,3,4} relative to {1,2,3} is {1}.
- Symmetric difference of sets A and B is the set whose members are members of exactly one of A and B. For instance, for the sets {1,2,3} and {2,3,4}, the symmetric difference set is {1,4}.
- Cartesian product of A and B, denoted $A \times B$, is the set whose members are all possible ordered pairs (a,b) where a is a member of A and b is a member of B.

The power set of a set A is the set whose members are all possible subsets of A. For example, the powerset of {1, 2} is { {}, {1}, {2}, {1,2} }.

Some Ontology

A set is pure if all of its members are sets, all members of its members are sets, and so on. For example, the set containing only the empty set is a nonempty pure set. When doing set theory, it is common to restrict attention to the pure sets, and many systems of axiomatic set theory are designed to axiomatize the pure sets only.

A key idea in set theory is the von Neumann universe of pure sets. Sets in this universe are arranged in a cumulative hierarchy, based on how deeply their members, members of members, etc. are nested. Each set is assigned an ordinal number á in this hierarchy, known as its rank.

A set is assigned a rank by transfinite recursion: if the least upper bound on the ranks of the members of a set X is α then X is assigned rank $\alpha + 1$. Also, for each ordinal á, the set V_α contains all sets assigned a rank no greater than α.

Axiomatic Set Theory

Elementary set theory can be studied informally and intuitively, and so can be taught in primary schools using, say, Venn diagrams. The intuitive approach silently assumes that all objects in the universe of discourse satisfying any defining condition form a set. This assumption gives rise to antinomies, the simplest and best known of which being Russell's paradox. Axiomatic set theory was originally devised to rid set theory of such antinomies.

The most widely studied systems of axiomatic set theory imply that all sets form a cumulative hierarchy. Such systems come in two flavours, those whose ontology consists of:

- *Sets alone.* This includes the most common axiomatic set theory, Zermelo–Fraenkel set theory (ZFC), which includes the axiom of choice. Fragments of ZFC include:
  - Zermelo set theory, which replaces the axiom schema of replacement with that of separation;
  - General set theory, a small fragment of Zermelo set theory sufficient for the Peano axioms and finite sets;
  - Kripke-Platek set theory, which omits the axioms of infinity, powerset, and choice, and weakens the axiom schemata of separation and replacement.
- *Sets and proper classes.* This includes Von Neumann-Bernays-Gödel set theory, which has the same strength as ZFC for theorems about sets alone, and Morse-Kelley set theory, which is stronger than ZFC.

For the above systems, allowing urelements (objects that can be members of sets while having no members themselves) does not give rise to any interesting mathematics.

The systems NFU (allowing urelements) and NF (lacking them), though having their origin in type theory, are not based on a cumulative hierarchy. NF and NFU include a "set of everything," relative to which every set has a complement. Here urelements matter, because NF, but not NFU, produces sets for which the axiom of choice does not hold.

Systems of constructive set theory, such as CST, CZF, and IZF, embed their set axioms in intuitionistic logic instead of first order logic. Yet other systems accept standard first order logic but feature a nonstandard membership relation. These include rough set theory and fuzzy set theory, in which the value of an atomic formula embodying the membership relation is not simply True and False. The Boolean-valued models of ZFC are a related subject.

Applications

Nearly all mathematical concepts are now defined formally in terms of sets and set theoretic concepts. For example, mathematical structures as diverse as graphs, manifolds, rings, and vector spaces are all defined as sets having various (axiomatic) properties. Equivalence and order relations are ubiquitous in mathematics, and the theory of relations is entirely grounded in set theory.

Set theory is also a promising foundational system for much of mathematics. Since the publication of the first volume of *Principia Mathematica*, it has been claimed that most or even all mathematical theorems can be derived using an aptly designed set of axioms for set theory, augmented with many definitions, using first or second order logic.

For example, properties of the natural and real numbers can be derived within set theory, as each number system can be identified with a set of equivalence classes under a suitable equivalence relation whose field is some infinite set.

Set theory as a foundation for mathematical analysis, topology, abstract algebra, and discrete mathematics is likewise uncontroversial; mathematicians accept that (in principle) theorems in these areas can be derived from the relevant definitions and the axioms of set theory. Few full derivations of complex mathematical theorems from set theory have been formally verified, however, because such formal derivations are often much longer than the natural language proofs mathematicians commonly present. One verification project, Metamath, includes derivations of more than 10,000 theorems starting from the ZFC axioms and using first order logic.

Areas of Study

Set theory is a major area of research in mathematics, with many interrelated subfields.

Combinatorial Set Theory

Combinatorial set theory concerns extensions of finite combinatorics to infinite sets. This includes the study of cardinal arithmetic and the study of extensions of Ramsey's theorem such as the Erdos-Rado theorem.

Descriptive Set Theory

Descriptive set theory is the study of subsets of the real line and, more generally, subsets of Polish spaces. It begins with the study of pointclasses in the Borel hierarchy and extends to the study of more complex hierarchies such as the projective hierarchy and the Wadge hierarchy. Many properties of Borel sets can be established in ZFC, but proving these properties hold for more complicated sets requires additional axioms related to determinacy and large cardinals.

The field of effective descriptive set theory is between set theory and recursion theory. It includes the study of lightface pointclasses, and is closely related to hyperarithmetical theory. In many cases, results of classical descriptive set theory have effective versions; in some cases, new results are obtained by proving the effective version first and then extending ("relativizing") it to make it more broadly applicable.

A recent area of research concerns Borel equivalence relations and more complicated definable equivalence relations. This has important applications to the study of invariants in many fields of mathematics.

Fuzzy Set Theory

In set theory as Cantor defined and Zermelo and Fraenkel axiomatized, an object is either a member of a set or not. In fuzzy set theory this condition was relaxed by Zadeh so an object has a *degree of membership* in a set, as number between 0 and 1. E.g. the degree of membership of a person in the set of "tall people" is more flexible than a simple yes or no answer and can be a real number such as 0.75.

Inner Model Theory

An inner model of Zermelo–Fraenkel set theory (ZF) is a transitive class that includes all the ordinals and satisfies all the axioms of ZF. The canonical example is the constructible universe L developed by Gödel. The study of inner models of extensions of ZF is of interest in set theory because it can be used to prove consistency results.

For example, it can be shown that regardless whether a model V of ZF satisfies the continuum hypothesis or the axiom of choice, the inner model L constructed inside the original model will satisfy both the generalized continuum hypothesis and the axiom of choice. Thus the assumption that ZF is consistent (has any model whatsoever) implies that ZF together with these two principles is consistent.

The study of inner models is common in the study of determinacy and large cardinals, especially when considering axioms such as the axiom of determinacy that contradict the axiom of choice. Even if a fixed model of set theory satisfies the axiom of choice, it is possible for an inner model to fail to satisfy the axiom of choice. For example, the existence of sufficiently large cardinals implies that there is an inner model satisfying the axiom of determinacy (and thus not satisfying the axiom of choice).

Large Cardinals

A large cardinal is a cardinal number with an extra property. Many such properties are studied, including inaccessible cardinals, measurable cardinals, and many more. These properties typically imply the cardinal number must be very large, with the existence of a cardinal with the specified property unprovable in Zermelo-Fraenkel set theory.

Determinacy

Determinacy refers to the fact that, under appropriate assumptions, certain two-player games of perfect information are determined from the start in the sense that one player must have a winning strategy. The existence of these strategies has important consequences in descriptive set theory, as the assumption that a broader class of games is determined often implies that a broader class of sets will have a topological property.

The axiom of determinacy (AD) is an important object of study; although incompatible with the axiom of choice, AD implies that all subsets of the real line are well behaved (in particular, measurable and with the perfect set property). AD can be used to prove that the Wadge degrees have an elegant structure.

Forcing

Paul Cohen invented the method of forcing while searching for a model of ZFC in which the axiom of choice or the continuum hypothesis fails. Forcing adjoins to some given model of set theory additional sets in order to create a larger model with properties determined (i.e. "forced") by the construction and the original model.

For example, Cohen's construction adjoins additional subsets of the natural numbers without changing any of the cardinal numbers of the original model. Forcing is also one of two methods for proving relative consistency by finitistic methods, the other method being Boolean-valued models.

Cardinal Invariants

A cardinal invariant is a property of the real line measured by a cardinal number. For example, a well-studied invariant is the smallest cardinality of a collection of meagre sets of reals whose union is the entire real line.

These are invariants in the sense that any two isomorphic models of set theory must give the same cardinal for each invariant. Many cardinal invariants have been studied, and the relationships between them are often complex and related to axioms of set theory.

Set-theoretic Topology

Set-theoretic topology studies questions of general topology that are set-theoretic in nature or that require advanced methods of set theory for their solution. Many of these theorems are independent of ZFC, requiring stronger axioms for their proof.

A famous problem is the normal Moore space question, a question in general topology that was the subject of intense research. The answer to the normal Moore space question was eventually proved to be independent of ZFC.

Objections to Set Theory

From set theory's inception, some mathematicians objected to it as a foundation for mathematics, arguing, for example, that it is just a game which included elements of fantasy. The most common objection to set theory, one Kronecker voiced in set theory's earliest years, starts from the constructivist view that mathematics is loosely related to computation.

If this view is granted, then the treatment of infinite sets, both in naive and in axiomatic set theory, introduces into mathematics methods and objects that are not computable even in principle. Ludwig Wittgenstein questioned the way Zermelo–Fraenkel set theory handled infinities. Wittgenstein's views about the foundations of mathematics were later criticised by Paul Bernays, and closely investigated by Crispin Wright, among others. In the mid 20th century, Errett Bishop dismissed set theory as "God's mathematics, which we should leave for God to do."

Category theorists have proposed topos theory as an alternative to traditional axiomatic set theory. Topos theory can interpret various alternatives to that theory, such as constructivism, finite set theory, and computable set theory.

A History of Set Theory

The history of set theory is rather different from the history of most other areas of mathematics. For most areas a long process can usually be traced in which ideas evolve until an ultimate flash of inspiration, often by a number of mathematicians almost simultaneously, produces a discovery of major importance.

Set theory however is rather different. It is the creation of one person, Georg Cantor. Before we take up the main story of Cantor's development of the theory, we first examine some early contributions.

The idea of infinity had been the subject of deep thought from the time of the Greeks. Zeno of Elea, in around 450 BC, with his problems on the infinite, made an early major contribution. By the Middle Ages discussion of the infinite had led to comparison of infinite sets.

For example Albert of Saxony, in *Questiones subtilissime in libros de celo et mundi,* proves that a beam of infinite length has the same volume as 3-space. He proves this by sawing the beam into imaginary pieces which he then assembles into successive concentric shells which fill space.

Bolzano was a philosopher and mathematician of great depth of thought. In 1847 he considered sets with the following definition *an embodiment of the idea or concept which we conceive when we regard the arrangement of its parts as a matter of indifference.*

Bolzano defended the concept of an infinite set. At this time many believed that infinite sets could not exist. Bolzano gave examples to show that, unlike for finite sets, the elements of an infinite set could be put in 1-1 correspondence with elements of one of its proper subsets. This idea eventually came to be used in the definition of a finite set.

It was with Cantor's work however that set theory came to be put on a proper mathematical basis. Cantor's early work was in number theory and he published a number of articles on this topic between 1867 and 1871. These, although of high quality, give no indication that they were written by a man about to change the whole course of mathematics.

An event of major importance occurred in 1872 when Cantor made a trip to Switzerland. There Cantor met Richard Dedekind and a friendship grew up that was to last for many years. Numerous letters between the two in the years 1873-1879 are preserved and although these discuss relatively little mathematics it is clear that Dedekind's deep abstract logical way of thinking was a major influence on Cantor as his ideas developed.

Cantor moved from number theory to papers on trigonometric series. These papers contain Cantor's first ideas on set theory and also important results on irrational numbers.Dedekind was working independently on irrational numbers and Dedekind published *Continuity and irrational numbers.*

In 1874 Cantor published an article in Crelle's Journal which marks the birth of set theory. A follow-up paper was submitted by Cantor to Crelle's Journal in 1878 but already set theory was becoming the centre of controversy. Kronecker, who was on the editorial staff of Crelle's Journal, was unhappy about the revolutionary new ideas contained in Cantor's paper. Cantor was tempted to withdraw the paper but Dedekind persuaded Cantor not to withdraw it and Weierstrass supported publication. The paper was published but Cantor never submitted any further work to Crelle's Journal. In his 1874 paper Cantor considers at least two different kinds of infinity. Before this orders of infinity did not exist but all infinite collections were considered 'the same size'. However Cantor examines the set of algebraic real numbers, that is the set of all real roots of equations of the form

$$a_n x^n + a_{n-1} x^{n-1} + a_{n-2} x^{n-2} + ... + a_1 x + a_0 = 0,$$

where a_i is an integer. Cantor proves that the algebraic real numbers are in one-one correspondence with the natural numbers in the following way.

For an equation of the above form define its index to be

$$|a_n| + |a_{n-1}| + |a_{n-2}| + ... + |a_1| + |a_0| + n.$$

There is only one equation of index 2, namely $x = 0$. There are 3 equations of index 3, namely

$$2x = 0,\ x + 1 = 0,\ x - 1 = 0 \text{ and } x^2 = 0.$$

These give roots 0, 1, -1. For each index there are only finitely many equations and so only finitely many roots. Putting them in 1-1 correspondence with the natural numbers is now clear but ordering them in order of index and increasing magnitude within each index.

In the same paper Cantor shows that the real numbers cannot be put into one-one correspondence with the natural numbers using an argument with nested intervals which is more complex than that used today (which is in fact due to Cantor in a later paper of 1891). Cantor now remarks that this proves a theorem due to Liouville, namely that there are infinitely many transcendental (i.e. not algebraic) numbers in each interval.

In his next paper, the one that Cantor had problems publishing in Crelle's Journal, Cantor introduces the idea of equivalence of sets and says two sets are equivalent or have the same power if they can be put in 1-1 correspondence. The word 'power' Cantor took from Steiner. He proves that the rational numbers have the smallest infinite power and also shows that R^n has the same power as R. He shows further that countably many copies of R still has the same power as R. At this stage Cantor does not use the word countable, but he was to introduce the word in a paper of 1883.

Cantor published a six part treatise on set theory from the years 1879 to 1884. This work appears in *Mathematische Annalen* and it was a brave move by the editor to publish the work despite a growing opposition to Cantor's ideas. The leading figure in the opposition was Kronecker who was an extremely influential figure in the world of mathematics.

Kronecker's criticism was built on the fact that he believed only in constructive mathematics. He only accepted mathematical objects that could be constructed finitely from the intuitively given set of natural numbers. When Lindemann proved that π is transcendental in 1882 Kronecker said:

> *Of what use is your beautiful investigation of p. Why study such problems when irrational numbers do not exist.*

Certainly Cantor's array of different infinities were impossible under this way of thinking.

Cantor however continued with his work. His fifth work in the six part treatise was published in 1883 and discusses well-ordered sets. Ordinal numbers are introduced as the order types of well-ordered sets. Multiplication and addition of transfinite numbers are also defined in this work although Cantor was to give a fuller account of transfinite arithmetic in later work. Cantor takes quite a portion of this article justifying his work. Cantor claimed that mathematics is quite free and any concepts may be introduced subject only to the condition that they are free of contradiction and defined in terms of previously accepted concepts. He also cites many previous authors who had given opinions on the concept of infinity including Aristotle, Descartes, Berkeley, Leibniz and Bolzano.

The year 1884 was one of crisis for Cantor. He was unhappy with his position at Halle and would have liked to move to Berlin. However this move was blocked by Schwarz and Kronecker. In 1884 Cantor wrote 52 letters to Mittag-Leffler each one of which attacked Kronecker. In this year of mental crisis Cantor seemed to lose confidence in his own work and applied to lecture on philosophy rather than on mathematics. The crisis did not last too long and by early 1885 Cantor was recovered and his faith in his own work had returned.

However, despite a wealth of important work in the years after 1884, there is some indication that he never quite reached the heights of genius that his remarkable papers showed over the 10 year period from 1874 to 1884.

Although not of major importance in the development of set theory it is worth noting that Peano introduced the symbol for 'is an element of' in 1889. It comes from the first letter if the Greek word meaning 'is'.

In 1885 Cantor continued to extend his theory of cardinal numbers and of order types. He extended his theory of order types so that now his previously defined ordinal numbers became a special case. In 1895 and 1897 Cantor published his final double treatise on sets theory. It contains an introduction that looks like a modern book on set theory, defining set, subset, etc. Cantor proves that if A and B are sets with A equivalent to a subset of B and B equivalent to a subset of A then A and B are equivalent. This theorem was also proved by Felix Bernstein and independently by E Schröder.

The dates 1895 and 1897 are important for set theory in another way. In 1897 the first published paradox appeared, published by Cesare Burali-Forti. Some of the impact of this paradox was lost since Burali-Forti got the definition of a well-ordered set wrong! However, even if the definition was corrected, the paradox remained. It basically revolves round the set of all ordinal numbers. The ordinal number of the set of all ordinals must be an ordinal and this leads to a contradiction. It is believed that Cantor discovered this paradox himself in 1885 and wrote to Hilbert about it in 1886. This is slightly surprising since Cantor was highly critical of the Burali-Forti paper when it appeared. The year 1897 was important for Cantor in another way, for in that year the first International Congress of Mathematicians was held in Zurich and at that conference Cantor's work was held in the highest esteem being praised by many including Hurwitz and Hadamard.

In 1899 Cantor discovered another paradox which arises from the set of all sets. What is the cardinal number of the set of all sets? Clearly it must be the greatest possible cardinal yet the cardinal of the set of all subsets of a set always has a greater cardinal than the set itself. It began to look as if the criticism of Kronecker might be at least partially right since extension of the set concept too far seemed to be producing the paradoxes. The 'ultimate' paradox was found by Russell in 1902 (and found independently by Zermelo). It simplify defined a set

$$A = \{ X \mid X \text{ is not a member of } X \}.$$

Russell then asked :- *Is A an element of A*? Both the assumption that A is a member of A and A is not a member of A lead to a contradiction. The set construction itself appears to give a paradox.

Russell wrote to Frege telling him about the paradox. Frege had been near completion of his major treatise on the foundations of arithmetic. Frege added an acknowledgement to his treatise.

> *A scientist can hardly meet with anything more undesirable than to have the foundation give way just as the work is finished. In this position I was put by a letter from Mr Bertrand Russell as the work was nearly through the press.*

By this stage, however, set theory was beginning to have a major impact on other areas of mathematics. Lebesgue defined 'measure' in 1901 and in 1902 defined the Lebesgue integral using set theoretic concepts. Analysis needed the set theory of Cantor, it could not afford to limit itself to intuitionist style mathematics in the spirit of Kronecker. Rather than dismiss set theory because of the paradoxes, ways were sought to keep the main features of set theory yet eliminate the paradoxes.

Did the paradoxes come from the 'Axiom of choice'? Cantor had used the 'Axiom of choice' without feeling that it was necessary to single it out for any special treatment. The first person to explicitly note that he was using such an axiom seems to have been Peano in 1890 in dealing with an existence proof for solutions to a system of differential equations. Again in 1902 it was mentioned by Beppo Levi but the first to formally introduce the axiom was Zermelo when he proved, in 1904, that every set can be well-ordered. This theorem had been conjectured by Cantor. Émile Borel pointed out that the Axiom of Choice is in fact equivalent to Zermelo's Theorem.

Gödel showed, in 1940, that the Axiom of Choice cannot be disproved using the other axioms of set theory. It was not until 1963 that Paul Cohen proved that the Axiom of Choice is independent of the other axioms of set theory.

Russell's paradox had *undermined the whole of mathematics* in Frege's words. Russell, trying to repair the damage, made an attempt to put mathematics back onto an logical basis in his major work *Principia Mathematica* written with Whitehead. This work attempts to reduce the foundations of mathematics to logic and was extremely influential. However the method of avoiding the paradoxes by introducing a 'theory of types' made it impossible to say that a class was or was not a member of itself. It did not seem a very satisfactory way around the problems and others sought different ways.

Zermelo in 1908 was the first to attempt an axiomatisation of set theory. Many other mathematicians attempted to axiomatise set theory. Fraenkel, von Neumann, Bernays and Gödel are all important figures in this development. Gödel showed the limitations of any axiomatic theory and the aims of many mathematicians such as Frege and Hilbert could never be achieved.

Analytic Function

In mathematics, an analytic function is a function that is locally given by a convergent power series. Analytic functions can be thought of as a bridge between polynomials and general functions. There exist both real analytic functions and complex analytic functions, categories that are similar in some ways, but different in others. Functions of each type are infinitely differentiable, but complex analytic functions exhibit properties that do not hold generally for real analytic functions. A function is analytic if and only if it is equal to its Taylor series in some neighborhood of every point.

Definitions

Formally, a function f is *real analytic* on an open set D in the real line if for any x_0 in D one can write

$$f(x) = \sum_{n=0}^{\infty} a_n (x - x_0)^n = a_0 + a_1(x - x_0) + a_2(x - x_0)^2 + a_3(x - x_0)^3 + \ldots$$

in which the coefficients a_0, a_1,... are real numbers and the series is convergent to f for x in a neighborhood of x_0.

Alternatively, an analytic function is an infinitely differentiable function such that the Taylor series at any point x_0 in its domain

$$T(x) = \sum_{n=0}^{\infty} \frac{f^{(n)}(x_0)}{n!}(x - x_0)^n$$

converges to $f(x)$ for x in a neighbourhood of x_0. The set of all real analytic functions on a given set D is often denoted by $C^{\grave{u}}(D)$.

A function f defined on some subset of the real line is said to be real analytic at a point x if there is a neighbourhood D of x on which f is real analytic.

The definition of a *complex analytic function* is obtained by replacing, in the definitions above, "real" with "complex" and "real line" with "complex plane."

Examples

Most special functions are analytic (at least in some range of the complex plane). Typical examples of analytic functions are:

- Any polynomial (real or complex) is an analytic function. This is because if a polynomial has degree n, any terms of degree larger than n in its Taylor series expansion will vanish, and so this series will be trivially convergent.
- The exponential function is analytic. Any Taylor series for this function converges not only for x close enough to x_0 (as in the definition) but for all values of x (real or complex).
- The trigonometric functions, logarithm, and the power functions are analytic on any open set of their domain.

Typical examples of functions that are not analytic are:

- The absolute value function when defined on the set of real numbers or complex numbers is not everywhere analytic because it is not differentiable at 0.

  Piecewise defined functions (functions given by different formulas in different regions) are typically not analytic where the pieces meet.
- The complex conjugate function is not complex analytic, although its restriction to the real line is real analytic.

Alternate Characterizations

If f is an infinitely differentiable function defined on an open set $D \subset \mathrm{R}$, then the following conditions are equivalent.

1) f is real analytic.
2) There is a complex analytic extension of f to an open set $G \subset \mathrm{C}$ which contains D.
3) For every compact set $K \subset D$ there exists a constant C such that for every $x \in K$ and every non-negative integer k the following estimate holds

$$\left|\frac{d^k f}{dx^k}(x)\right| \leq C^{k+1} k!$$

The real analyticity of a function f at a given point x can be characterized using the FBI transform.

Complex analytic functions are exactly equivalent to holomorphic functions, and are thus much more easily characterized.

Properties of Analytic Functions

- The sums, products, and compositions of analytic functions are analytic.
- The reciprocal of an analytic function that is nowhere zero is analytic, as is the inverse of an invertible analytic function whose derivative is nowhere zero.
- Any analytic function is smooth, that is, infinitely differentiable. The converse is not true; in fact, in a certain sense, the analytic functions are sparse compared to all infinitely differentiable functions.
- For any open set $\cap \subseteq C$, the set $A(\cap)$ of all bounded, analytic functions $u : \cap \rightarrow C$ is a Banach space with respect to the supremum norm. The fact that uniform limits of analytic functions are analytic is an easy consequence of Morera's theorem.

A polynomial cannot be zero at too many points unless it is the zero polynomial (more precisely, the number of zeros is at most the degree of the polynomial). A similar but weaker statement holds for analytic functions. If the set of zeros of an analytic function f has an accumulation point inside its domain, then f is zero everywhere on the connected component containing the accumulation point.

More formally this can be stated as follows. If (r_n) is a sequence of distinct numbers such that $f(r_n) = 0$ for all n and this sequence converges to a point r in the domain of D, then f is identically zero on the connected component of D containing r.

Also, if all the derivatives of an analytic function at a point are zero, the function is constant on the corresponding connected component. These statements imply that while analytic functions do have more degrees of freedom than polynomials, they are still quite rigid.

Analyticity and Differentiability

As noted above, any analytic function (real or complex) is infinitely differentiable (also known as smooth, or C^{∞}). There exist smooth real functions which are not analytic: see non-analytic smooth function. In fact there are many such functions, and the space of real analytic functions is a proper subspace of the space of smooth functions.

The situation is quite different when one considers complex analytic functions and complex derivatives. It can be proved that any complex function differentiable (in the complex sense) in an open set is analytic. Consequently, in complex analysis, the term *analytic function* is synonymous with *holomorphic function.*

Real versus Complex Analytic Functions

Real and complex analytic functions have important differences (one could notice that even from their different relationship with differentiability). Analyticity of complex functions is a more restrictive property, as it has more restrictive necessary conditions and complex analytic functions have more structure than their real-line counterparts.

According to Liouville's theorem, any bounded complex analytic function defined on the whole complex plane is constant. This statement is clearly false for real analytic functions, as illustrated by

$$f(x) = \frac{1}{x^2 + 1}.$$

Also, if a complex analytic function is defined in an open ball around a point x_0, its power series expansion at x_0 is convergent in the whole ball. This is not true in general for real analytic functions.

Any real analytic function on some open set on the real line can be extended to a complex analytic function on some open set of the complex plane. However, not every real analytic function defined on the whole real line can be extended to a complex function defined on the whole complex plane. The function $f(x)$ defined in the paragraph above is a counterexample, as it is not defined for $x = \pm i$.

Analytic Functions of Several Variables

One can define analytic functions in several variables by means of power series in those variables. Analytic functions of several variables have some of the same properties as analytic functions of one variable. However, especially for complex analytic functions, new and interesting phenomena show up when working in 2 or more dimensions. For instance, zero sets of complex analytic functions in more than one variable are never discrete.

Working With Expressions

One of the important algebraic procedures you will be called upon to do is *evaluate expressions.* That means reducing the expression to a numeric value, or its simplified form. Evaluating expressions is not always simple. There is a systematic way to evaluate expressions to ensure you get the correct answer. If you are given an expression that contains a variety of different operations, there is a specific order in which these operations must be performed. This order is called the *order of operations.*

In this unit, we will review the notation, concepts, and procedures required for evaluating expressions. Since it is critical to understand exponents in order to understand later procedures, we will start there. We will follow with multiplication, division, addition, and subtraction of algebraic terms, and then discuss the use of parentheses and other grouping symbols. We will then apply all of these as we review the order of operations.

Exponents

Exponential notation or *exponential form* is commonly found in algebraic terms, expressions, and equations. Exponents are used to shorten or condense repeated multiplication. For example, a term containing an exponent is shown below:

$$7^4$$

In this term, 7 is the *base* and 4 is the *exponent.* The *exponent* or *power,* indicates the number of times that the factor, or *base,* is multiplied.

$$7^4 = 7 \cdot 7 \cdot 7 \cdot 7$$

When working with expressions containing only numbers, simply perform the indicated multiplication. Below are three examples:

$15^2 = 15 \cdot 15 = 225$ $\quad$ $3^1 = 3$ $\quad$ $2^7 = 2 \cdot 2 \cdot 2 \cdot 2 \cdot 2 \cdot 2 \cdot 2 = 128$

Notice in the second example that a number raised to the first power is just that number ($3^1 = 3$). When you see a number that does not have an exponent, it is because the 1 is assumed. When variables are included in the terms, we work according to the same principles.

Exponents in Algebra

Exponents included in algebraic terms and expressions can apply either to just one *variable,* an entire *term,* or an *expression.* When a *term* or *expression* is raised to a power, we must apply the exponent to the entire term or collection of terms.

A Variable Raised to a Power

If the term is simply one variable, the exponent is handled just as it was in the previous section. If we want to raise one number or variable to a power we put the exponent as a superscript to that one number or variable. An algebraic term containing an exponent is shown below.

In this term, x is the base, and 2 is the exponent. The *exponent,* or *power*, indicates the number of times that the factor, or *base,* occurs.

$$x^2 = x \cdot x$$

In the example at the right, the term contains the variable to the fourth power. The numerical coefficient is *not* raised to the power. y is the base, and 4 is the exponent.

$$3y^4 = 3 \cdot y \cdot y \cdot y \cdot y$$

As with numbers, a variable that does not appear to have a power is raised the first power.

$$z^1 = z$$

A Term Raised to a Power

Remember, a *term* is either a number or a product of a number and one or more variables.

This includes any powers to which the variables are raised. In a term, either the entire term, or just one variable within the term, may be raised to a power.

To indicate that the entire term is raised to a power, we must enclose the term in parentheses and place the exponent outside the parentheses. The general rule for applying an exponent to a whole term contained within parentheses is:

$$(x^a)^b = x^{a \cdot b}$$

The result is much different than when the parentheses are omitted. In the example at the right, the entire term is raised to the second power; this is commonly referred to as squared. Notice the parentheses around the entire term.

$$(5x)^2 = 5^{1 \cdot 2}x^{1 \cdot 2} = 5^2x^2 = 5 \cdot 5 \cdot x \cdot x = 25x^2$$

Let's contrast the above example to the one at the right. In this example there are no parentheses grouping 5 and x. This means only the variable x is squared.

$$5x^2 = 5 \cdot x \cdot x$$

If we have a fraction inside parentheses raised to a power, the same rules apply. Each number and variable inside the parentheses must be raised to the power.

$$(2x^3 / y)^2 = 2^{1\cdot 2}x^{3\cdot 2} / y^{1\cdot 2} = 2^2x^6 / y^2 = 4x^6 / y^2$$

Multiplying Two Terms with the Same Variable

At times we must multiply together variables that are raised to different exponents. When multiplying together these types of terms, you add the exponents of each of the variables.

$$x^c \bullet x^d = x^{c+d}$$

At first it may not be clear why this should be done. Look at an example of $x2 \cdot x^4$. If we expand both of these terms we have:

$$x^2 = x \cdot x \text{ and } x^4 = x \cdot x \cdot x \cdot x$$

If we multiply these together we have:

$$x2 \cdot x^4 = x \cdot x \cdot x \cdot x \cdot x \cdot x = x^6 = x^{2+4}$$

General Rules for Using Exponents with Variables and Terms

- Any number or variable that appears to have no exponent is raised to the first power.

  $x^1 = x$

- If you raise any number or variable to the power of zero, the result is one.

  $5^0 = 1$, and $z^0 = 1$

- To apply an exponent to a term in exponential form, multiply the exponents.

  $(y^d x^a)^b = y^{d\cdot b}x^{a\cdot b}$

- An exponent outside the parentheses applies to all parts of a product or quotient inside the parentheses:

  $(xy)^a = (x \bullet y)^a = x^a \bullet y^a$ and $(x/y)^a = x^a/y^a$

- When multiplying together two terms with the same variables, add the exponents.

  $x^c \bullet x^d = x^{c+d}$

- A number or variable to a negative power means to move it to the denominator of a fraction and put it to the positive power.

  $5^{-3} = 1/5^3$, and $z^{-2} = 1/z^2$

Example

Multiply the following terms.

1. $(4x^3y)^3$
2. $(3x^2/5y^2)^4$

Answers: First multiply the numerical coefficients, then use the exponent rules.

1. $(4x^3y)^3 = 4^3x^{3\cdot 3}y^{1\cdot 3} = 64x^9y^3$
2. $(3x^2/5y^2)^4 = 3^{1\cdot 4}x^{2\cdot 4}/5^{1\cdot 4}y^{2\cdot 4} = 81x^8/625y^8$

Exponents in Algebraic Expressions

If we have a collection of terms, called an expression, which is raised to an exponent, we again must apply the exponent to the entire expression. We place parentheses (or another grouping symbol) around the expression. The exponent is then placed outside the parentheses. Below is an example of an expression raised to a power:

$$(15x^2 + xy)^3 = (15x^2 + xy)\ (15x^2 + xy)\ (15x^2 + xy)$$

The next section will apply some of the basic rules of exponents used when multiplying and dividing algebraic terms.

Multiplying and Dividing Algebraic Terms

Multiplying Algebraic Terms

When multiplying algebraic terms you (1) multiply numerical coefficients together, then (2) list all the variables that occur in the terms being multiplied and (3) add the exponents of like variables.

When doing this, we can use what we learned in the last section on exponents. Below are some examples of multiplication applied to algebraic terms.

In this example, the base of both terms is x. This means we only need to multiply the coefficient for each term, attach the variable, then add the exponents. Notice that when no exponent is explicitly stated, the exponent is assumed to be one.

$$x^2 \cdot 3x = 1 \cdot 3 \cdot x^{2+1} = 3x^3$$

Here, we multiply the numerical coefficients, keep the variable and add the exponents of like variables.

$$5k^3 \cdot k^2 = 5 \cdot 1 \cdot k^{3+2} = 5k^5$$

Both examples above contained terms with the same variables in their bases. However, even when the bases are not the same, we follow the same procedure. We multiply the numerical coefficients and then simply list the variables from each term.

For example, if we multiply the terms $3xy$ and $5y^2$, we get the result $15xy^3$. This works because we can view $5y^2$ as $5x^0y^2$. (NOTE: In the previous section we mentioned that $x^0 = 1$.)

So we can write this multiplication as:

$$3xy \cdot 5y^2 = 3x^1y^1 \cdot 5x^0y^2 = 3 \cdot 5 \cdot x^{1+0} \cdot y^{1+2} = 15xy^3$$

Example: Multiply the following terms.

1. $2x^2y \cdot 6x^2y$
2. $3ab \cdot 5xy^2$

Answers: First multiply the numerical coefficients, then use the exponent rules.

1. $2x^2y \cdot 6x^2y = (2 \cdot 6)(x^2y \cdot x^2y) = 12x^{2+2}y^{1+1} = 12x^4y^2$
2. $3ab \cdot 5xy^2 = (3 \cdot 5)abxy^2 = 15abxy^2$

Dividing Algebraic Terms

The process for division of algebraic expressions is much like the multiplication process, with the exception that you divide the numerical coefficients and then subtract exponents instead of adding.

The exponent rules for division are:

Exponent Rules for Division

- If there are numerical coefficients in the expressions to be divided, just divide the numerical coefficient and then use the exponent rule to divide the variables.

$$\frac{4x^a}{2x^b} = \frac{4}{2}x^{a-b} = 2x^{a-b}$$

- To divide variables, keep the variables and subtract the exponents.

$$\frac{x^a y^c z}{x^b y^d} = x^{a-b} y^{c-d} z$$

If we divide $12xy^2$ by $3y$ we get $4xy$ as a result. This works because we can look at the term in the bottom of the fraction as being $3x^0y^1$.

$$\frac{12xy^2}{3y}=\frac{12}{3}x^{1-0}y^{2-1}=4x^1y^1=4xy$$

This subtraction of exponents looks as if we are cancelling out numbers and variables.

This is often expressed in the way shown below. (If you feel you need to review how to work with fractions, you should return to Book I of this series, Review of Fractions.)

$$\frac{12xy^2}{3y}=\frac{\not{3}\bullet 4\bullet x\bullet y\bullet y}{\not{3}\bullet y}=4xy$$

Because we subtract exponents, it is possible to get a negative exponent.

NOTE: When we get an answer in this form, with negative exponents, we may leave them in this form, or rewrite the expression according to the following rule.

$x^{-n} = 1/x^n$

So if we have divide $2kr$ by kr^4 we get:

$$\frac{2kr}{kr^4}=\frac{2}{1}k^{1-1}r^{1-4}=2k^0r^{-3}=2r^{-3}=\frac{2}{r^3}$$

Now let's look at an example of division problems.

Example: Divide the following terms.

1. b^6, b^3
2. k^4z^9/k^2z^5
3. $20a^4b^5 / 5a^7b^4$

Answer: *Divide the following terms.*

1. $b^6, b^3 = b^6/b^3 = b^{6-3} = b^3$
2. $k^4z^9/ k^2z^5 = k^{4-2}z^{9-5} = k^2z^4$
3. $20a^4b^5 / 5a^7b^4 = (20 \div 5)a^{4-7}b^{5-4} = 4a^{-3}b$

This solution can be rewritten as $4a^{-3}b = 4b/a^3$

Multiplying and Dividing Algebraic Terms

Below is a brief outline of the steps for multiplying and dividing terms that contain variables with exponents.

Multiplying Terms

1. Multiply the numerical coefficients. (Do not confuse the numerical coefficients with the base.)
2. Perform the multiplication on the variables:
   - o Keep the variables from each term.
   - o For each variable, add the exponents that appear for each variable.

     $2z^a x^d \cdot 3z^c x^b = (2 \cdot 3)z^{a+c} x^{d+b+b} = 6\ z^{a+c} x^{d+b}$

Dividing Terms

1. Divide the numerical coefficients.
2. Perform the division on the variables:
   - o Keep the variables from each term.
   - o For each variable, subtract the exponent on the bottom of the fraction from the exponent on the top of the fraction.

     $6x^c y^a / 2xy^b = (6/2)x^{c-1} y^{a-b} = 3x^{c-1} y^{a-b}$

Adding and Subtracting Like Terms

There is a very simple property for adding and subtracting algebraic expressions. To be able to add or subtract expressions, we must have *like terms. Like terms* are terms that contain the same variable or group of variables raised to the same exponent, regardless of their numerical coefficient.

For example:

$3x$ and $6x$ are *like terms.* They both contain x.

$6c^2$ and $19c^2$ are *like terms.* They both contain c^2.

$2xy^3$ and $101xy^3$ are *like terms.* They both contain xy^3.

$-km^2x^5$ and $17km^2x^5$ are *like terms.* They both contain km^2x^5.

Notice that to determine *like terms*, you must consider the variables in each term as a group. *Like terms* are those with exactly the same variables raised to the same exponent.

If two terms have the same variables, but to different powers, they are not *like terms* and cannot be combined.

For example:

x^4 and $3x^2$ are not like terms since one contains x^4 and the other contains x^2 as variables.

$5vk^3$ and vk^2 are not like terms since one contains vk^3 and the other contains vk^2 as variables

Adding and Subtracting Like Terms

To combine *like terms*, do the following:

1. Determine which terms contain the same variable or groups of variables raised to the same exponent.
2. Add or subtract the numerical coefficients.
3. Attach the common variables and exponents.

For example, $3x + 6x$ can be simplified to $(3 + 6)\,x = 9x$

Example

If possible, simplify each of the following expressions.

1. $10a + 10b - 3a$
2. $5b^2 + 8b^3$
3. $3x^2y^2z - 5xyz + x^2y^2z$

Answers

1. $10a + 10b - 3a = 7a + 0b$
2. $5b^2 + 8b^3 = 5b^2 + 8b^3$
3. $3x^2y^2z - 5xyz + x^2y^2z = 4x^2y^2z - 5xyz$

Detailed Answers

(1) $10a + 10b - 3a = 7a + 10b$

To find the solution:

1. Determine which terms contain the same variable or groups of variables raised to the same exponent: If we look at this equation we see that there are two terms which contain the variable a.

   $10a + 10b - 3a$

2. Add or subtract the numerical coefficients: We use the distributive property to rewrite the equation. Then perform the subtraction indicated, subtracting 3 from 10.

   $(10 - 3)\,a + 10b$

3. Attach the common variables and exponents: We then display the final result from the subtraction.

   $7a + 10b$

   (2) $5b^2 + 8b^3 = 5b^2 + 8b^3$

Determine which terms contain the same variable or groups of variables raised to the same exponent: While both terms have *b*'s in them, they are raised to different powers, b^2 and b^3.

This means we cannot combine these two terms.

$5b^2 + 8b^3$

(3) $3x^2y^2z - 5xyz + x^2y^2z = 4x^2y^2z - 5xyz$

Determine which terms contain the same variable or groups of variables raised to the same exponent: If we look at this equation we see that there are two terms which contain the variable :

x^2y^2z. $3x^2y^2z - 5xyz + x^2y^2z$

Add or subtract the numerical coefficients: We group these terms together and perform the addition indicated, adding 3 and 1 together.

$(3 + 1)(x^2y^2z) - 5xyz$

Attach the common variables and exponents: Here we have the final result.

$4x^2y^2z - 5xyz$

Combining like terms is crucial in solving equations. This is a procedure you will use often with algebraic expressions.

Use of Parentheses and Other Grouping Symbols

Already in this unit you have seen the use of parentheses. In the section on exponents we used parentheses to illustrate how entire terms were raised to a power.

This was an example of using parentheses for grouping. Using grouping symbols helps us to identify the order in which we should apply mathematical operations.

When we have more complex expressions, which may combine several terms, and use multiple operations, we may need to group terms to help stay organized. Parentheses, ( ), are most commonly used in grouping but you may also see brackets, [ ], or braces, { }.

When a term or expression is inside one of these grouping symbols it means that any operation indicated to be done on the group is done to the entire term or expression.

For example, in the section on exponents, when a term inside parentheses is raised to a power, it means we raise the entire term to that power.

With terms, this means we raise each numerical coefficient and variable in the term to that power.

$$(2x)^2 = (2x)(2x) = 2^2x^2 = 4x^2$$

Also, when we raise an expression with parentheses to a power, it means to multiply the entire expression by itself the number of times indicated by the exponent.

$$(5x^2 + y)^4 = (5x^2 + y)\ (5x^2 + y)\ (5x^2 + y)\ (5x^2 + y)$$

When there are multiple grouping symbols in a single expression, the preferred way to write them is {[( )]} with the parentheses on the inside, the square-shaped brackets used next, and braces used outermost.

An example of an expression that uses multiple grouping symbols is:

$$3\{[2(2x + 6) + 3x] - 12z\} = 21x + 36 - 36z$$

Grouping Algebraic Terms

When simplifying an expression, we always work from within the grouping symbols first. When we have multiple grouping symbols, we work from the innermost set towards the outside.

Below is an example of this.

- We first begin simplifying within the parentheses. In the example, this means adding the terms inside the parentheses.

  $3\ (5x + 2x) = 3\ (7x)$

- Then multiply the term inside the parentheses with the one outside the parentheses.

  $3\ (7x) = 21x$

The example above only has one set of grouping symbols, the parentheses. In the example below we have two sets of grouping symbols, parentheses and brackets.

We begin by performing the operation in the innermost grouping symbols, the parentheses.

$$15x\ [(3xy + 4xy) + 10xy] = 15x\ [7xy + 10xy]$$

We then perform the operation in the brackets.

$$15x\ [7xy + 10xy] = 15x\ [17xy]$$

Finally we perform the multiplication of the term outside the brackets with the one inside.

$$15x\ [17xy] = 255x^2y.$$

Order of Operations

The mathematical tools we've reviewed in this unit will help you to *evaluate expressions*. Remember, that means reducing an expression to its simplified form. The final tool you will need to evaluate complex expressions is the order of operations.

The order in which you perform these steps makes a difference. There is a systematic way to evaluate expressions to ensure you do it correctly.

For example, suppose you want to evaluate a simple expression. Look at the example shown below.

If we follow the order of operations, we get a result of 16.

$$6 + 2 \cdot 5 = 6 + 10 = 16$$

When dealing with algebraic expressions, or terms which contain variables, we follow the same order of operations. It is important to begin within the parentheses (or other grouping symbols) and follow the rest of the steps from there.

Example

Evaluate the following expressions.

1. $[(10 \cdot 2 + 8) \div 14]^2$
2. $12x^2y \div 4x + (6x - x)^2$
3. $(x + z) + (5x \cdot 2)$
4. $\{4\ [2xz^4\ (2c^3v \div 2cv^2)^3 \div 2]\} \div c^6x \cdot v^3$

Answers

Evaluate the following expressions.

1. $[(10 \cdot 2 + 8) \div 14]^2 = 4$
2. $12x^2y \div 4x + (6x - x)^2 = 3xy + 25x^2$
3. $(x + z) + (5x \cdot 2) = 11x + z$
4. $\{4\ [2xz^4\ (2c^3v \div 2cv^2)^3 \div 2]\} \div c^6x \cdot v^3 = 4z^4$

Detailed Answers

When evaluating expressions we first must consider the order of operations, PE[MD][AS].

1. $[(10 \cdot 2 + 8) \div 14]^2 = 4$

In this expression we have two sets of parentheses.

We begin by evaluating the expression in the innermost parentheses, following the order of operations within that inner set of parentheses. This is shown on the right.

$$[(10 \cdot 2 + 8) \div 14]^2 = [(20 + 8) \div 14]^2 = [28 \div 14]^2$$

We then look at the second expression in brackets and follow the order of operations within this set of brackets. This is shown on the right.

$$[28 \div 14]2 = [2]^2$$

Now we move on to exponents outside the brackets.

$$[2]^2 = 4$$

2. $12x^2y \div 4x + (6x - x)^2 = 3xy + 25x^2$

For this expression we work through the order of operations: P-Parentheses, E-Exponents, [M-Multiplication and D-Division], [A-Addition and S-Subtraction].

We must begin with the parentheses, $(6x - x)$, in this expression. We first do the subtraction here.

Because the terms are like terms, x can be subtracted from 6x. Once this is done, we perform the operation indicated by the exponent associated with the parentheses.

$$12x^2y \div 4x + (6x - x)^2 = 12x^2y \div 4x + (5x)^2 = 12x^2y \div 4x + 25x^2$$

Now we need to perform the multiplication and division in the expression. Notice that there is only division, no multiplication. When we divide $12x^2y$ by 4x, we get 3xy.

The next step is to perform any addition or subtraction. Since the two terms left in the expression are not like terms, they cannot be subtracted. The expression has been simplified.

$$12x^2y \div 4x + 25x^2 = 3xy + 25x^2$$

3. $(x + z) + (5x \cdot 2) = 11x + z$

We begin with both sets of parentheses and perform the indicated operations within each set of parentheses. Note that x and z in the first set are not like terms so they cannot be added.

$$(x + z) + (5x \cdot 2) = (x + z) + 10x$$

If we refer to the commutative property:

$$a + b = b + a,$$

and associative property:

$$a + (b + c) = (a + b) + c$$

we reorder and regroup the terms.

$$(x + z) + 10x = 10x + (x + z)$$

$$10x + (x + z) = (10x + x) + z$$

$$(10x + x) + z = 11x + z$$

4. $\{4\,[2xz^4\,(2c^3v \div 2cv^2)^3 \div 2]\} \div c^6x \cdot v^3 = 4z^4$

Here we have a number of sets of parentheses and brackets that group parts of the expression. We perform the order of operations from the inside out.

We begin with the inside set of parentheses, which contains division.

$$\{4\,[2xz^4\,(2c^3v \div 2cv^2)^3 \div 2]\} \div c^6x \cdot v^3 = \{4\,[2xz^4\,(c^2/v)^3 \div 2]\} \div c^6x \cdot v^3$$

Then the exponent outside of the parentheses.

$$\{4\,[2xz^4\,(c^2/v)^3 \div 2]\} \div c^6x \cdot v^3 = \{4\,[2xz^4\,(c^6/v^3) \div 2]\} \div c^6x \cdot v^3$$

Now we look at the inner set of brackets and follow the order of operations within this set of brackets. First we multiply $2xz^4$ by c^6/v^3.

$$\{4\,[2xz^4\,(c^6/v^3) \div 2]\} \div c^6x \cdot v^3 = \{4\,[2xz^4c^6/v^3 \div 2]\} \div c^6x \cdot v^3$$

Then we divide the result of this multiplication, $2xz^4c^6/v^3$, by 2.

$$\{4\,[2xz^4c^6/v^3 \div 2]\} \div c^6x \cdot v^3 = \{4\,[xz^4c^6/v^3]\} \div c^6x \cdot v^3$$

If we then look at the remaining braces, there is just a multiplication by 4 and division by v^3.

$$\{4\,[xz^4c^6/v^3]\} \div c^6x \cdot v^3 = 4xz^4c^6/v^3 \div c^6x \cdot v^3$$

We are now left only with an expression with no grouping symbols, so follow the remainder of the order of operations.

The first term in this expression is the fraction $4xz^4c^6/v^3$. We can rewrite the exponents of the variable in the denominator of this expression so it can be rewritten as $4xz^4c^6v^{-3}$.

We begin with the division of $4xz^4c^6v^{-3}$ by c^6x.

$$4xz^4c^6v^{-3} \div c^6x \cdot v^3 = 4x^{1-1}z^4c^{6-6}v^{-3} \cdot v^3 = 4z^4v^{-3} \cdot v^3$$

Then we need to multiply $4z^4v^{-3}$ by v^3.

$$4z^4v^{-3} \cdot v^3 = 4z^4v^{-3+3} = 4z^4$$

For each set of grouping symbols, the order of operations holds.

Functions and their Graphs

Graphs and Graphing Utilities

There are several kinds of graphs that the TI-82 or TI-83 will do, but they are accessed from the same key. The [y=] key and the [2nd] [y=] or [StatPlot] key.

The two major kinds of plots may cause confusion when mixed, so it is best to turn off all unnecessary plots before trying to graph an equation.

To turn off all statistical plots, hit and choose option 4 - PlotsOff. Hit enter to finish the command.

To turn off a regular plot, one can either hit while on an equation, or arrow to the left so that the cursor is on the equal sign, and press enter to toggle the display of that equation.

Graphing Equations with the Calculator

Equations must be solved explicitly for y before being placed into the calculator. In other words, y must be written as a function of x before entering the equation into the calculator.

Sometimes, equations can't be solved for y. Other times, the result is not a function. In this case, you may need two y functions.

$x^2 + y^2 = 16$ has the solutions y = + sqrt($16–x^2$) and y = –sqrt ($16 –x^2$)

When you solve the above equation, you get two values for y. Put the positive square root in $y_1$and the negative square root in y_2. Alternatively, you can specify that $y_2= –y_1$. On the TI-82, the y_1 variable can be found under [2nd] [Vars] (or [Y-Vars]), Function. On the TI-83, it is found under [Vars], Y–Vars, Function.

Zoom Settings

There are several zoom settings that we will be using in this class.

Standard

The standard zoom setting sets the domain and range to [-10,10] and the scale to 1. If you're not familiar with the interval notation [–10,10].

When moving the cursor in the standard setting, the x-step is 0.21276596 and the y-step is 0.32258065. These are not "nice" values, and often cause problems when trying to find exact values by tracing.

They also cause problems with the graph when there are vertical asymptotes (rational expressions) or at the endpoints of the domain (as was illustrated in class with the circle).

Decimal

The decimal zoom setting sets up the screen so that the x-step is 0.1 and the y-step is 0.1. This makes the values much nicer to look at, but gives you a much smaller window.

The domain for the decimal setting is [–4.7,4.7] and the range is [-3.1,3.1]. This also helps us figure out the size of the screen. If each pixel (dot on the screen) is 0.1 apart, then there are 95 pixels horizontally and 63 pixels vertically.

Square

Because the size of the screen is not the same, graphs have a tendency to not come out looking right. The square zoom setting is a way to remedy that.

In the decimal setting, the horizontal and vertical steps are both 0.1, so the images will appear square, but in the standard setting, the horizontal and vertical steps are 0.213 and 0.323 respectively.

The square decimal leaves the larger interval of the domain or range alone and changes the settings on the smaller one. If the domain and range are the same length (as in a standard setting), the domain is changed.

Zoom Out / Zoom In

If you want to maintain a square setting, but you need a larger portion of the viewing window to be shown, you can use the Zoom Out or Zoom In settings. Each of these require a position to zoom in or out about. So, after you choose the Zoom In or Zoom Out, you will need to position the cursor to where you want the centre of the zoom to be and press enter.

The zoom factor is four. That is, if you zoom out, you will see sixteen times as much area (four times the horizontal and four times the vertical makes sixteen times the area). As an example, assume your domain is [-10,10] and your range is [-5,5]. If you zoom out with the origin as the centre of the zoom, your new domain will be [-40,40] and hte new range will be [-20,20]. When zooming in, the factor is one-fourth. In other words, a zoom in done after a zoom out (with the same centre for both) will return to the previous settings.

Box

The zoom box option is like the zoom in option, except that you specify two corners of a box, and then the area between those corners is displayed.

Stat

The zoom statistics setting is useful when you have a statistical plot that you want to include all the data for.

The plot will set the Xmin and Xmax so that all of the data is visible. If you are graphing a scatter plot, so that there are both x and y data values, then the Ymin and Ymax will also be changed so that all of the data is visible.

Trig

Sets the domain to be from -2π to $+2\pi$ with the scale to be $\pi/2$. The range is [–4,4] with a scale of 1. This is a nice setting to use when graphing trigonometric functions, but we won't be doing that in this class, so you can pretty much ignore it.

Determining the Viewing Window

When you're trying to determine what the viewing window is, you should put the equation into the calculator so that you will be able to see when the graph matches the picture. Take into account the number of tick marks and the min and max settings.

The first thing to do is try to find some important points on the graph like y-intercepts, x-intercepts, or maximums and minimums.

If you can figure out where it's at on the graph from the calculator, you can then figure out what the scale is (how far apart the ticks are). Once you have that, it's pretty easy to find the viewing window by just counting the number of ticks to each side of zero.

Lines in the Plane

Slope of a Line

The slope is represented by the letter m.

The slope of a non-vertical line is defined several ways. It is the rise over the run. It is the change in y over the change in x.

For two points (x_1,y_1) and (x_2,y_2) where $x_1 \neq x_2$, the slope is $m = (y_2 - y_1)/ (x_2 - x_1)$

If the slope is positive, the line rises from left to right. If the slope is negative, the line falls from left to right. If the slope is zero, the line is horizontal. If the slope is undefined, the line is vertical.

Point-Slope Form of a Line

The equation of the non-vertical line passing through the points (x_1,y_1) and (x_2,y_2) and having slope m is given by the equation:

$$y - y_1 = m (x - x_1)$$

Which point you call point 1 and which point you call point 2 does not matter.

We almost never leave the equation of a line in point-slope form, but use it as a stepping ground to a final answer. One exception to this is when we're finding the asymptotes to hyperbolas in conic sections.

Historical Note

Linear Interpolation is using the equation of a line to approximate a value which falls between two known points. Linear Interpolation is often used when looking up values in a table, and the value you need is not in the table, but between two values which are in the table. In the good old days before there were calculators, we used linear interpolation to find logarithms and trigonometric values. Now, the calculators have those functions built into them, so there is less need to use interpolation. Linear Interpolation is still used some in statistics, but for most things, we just use the value from the table which is closer or less likely to cause a more serious error (don't worry if that didn't make sense, take a statistics course, and it will).

Linear Extrapolation is the process of using the equation of a line to approximate a value which falls outside two known points.

Slope-Intercept Form of a Line

The equation of the non-vertical line crossing the y-axis at the point (0,b) and having slope m is given by the equation:

$$y = m x + b$$

The point-slope form can be placed into the slope-intercept form with a little algebra.

The slope-intercept form of a line is what must be placed into the calculator to get it to graph the line.

General Form of a Line

In general, the general form of anything will be the form where all the variables and constants are on the left side of the equation, in decreasing degree of the terms and alphabetically for those terms that have the same degree.

For a line, that means $ax + by + c = 0$

a and b can not both be zero, if they were, then you would have $c=0$, which is a constant, not a linear function.

Vertical and Horizontal Lines

Vertical lines are lines that have all the x-coordinates the same. So, the equation of a vertical line is $x=a$ (where a is that common abscissa).

Horizontal lines are lines that have all the y-coordinates the same. So, the equation of a horizontal line is $y=b$ (where b is that common ordinate).

Parallel Lines

Parallel lines are lines in the same plane that do not intersect. The slope of parallel lines is the same.

Perpendicular Lines

Perpendicular lines are lines in the same plane that intersect at a right angle. The product of non-vertical and non-horizontal perpendicular lines is negative one. Another way of saying that is that the slopes of perpendicular lines are opposite reciprocals of each other.

Calculator

Using the graphing calculator to graph lines is pretty straightforward. Solve for y and enter the expression into the calculator. Solving for y is equivalent to putting the equation into slope-intercept form.

One warning though for users of the TI82 calculator. Be careful with slopes that are fractions.

Consider the following equations.

- $y = 1/2x + 2$
- $y = x/2 + 2$
- $y = (1/2)x + 2$

- y = 1/2*x + 2
- y = 0.5x + 2
- y = 0.5*x + 2.

All of the above *except* for the first one will give you the graph of a line. The first one will give you the graph of a rational function.

The TI82 *incorrectly* gives *implied* multiplication a higher preference than division and assumes that when people put an expression without a multiplication symbol after a division symbol, they wish to have the whole expression in the denominator.

The TI–82 takes y=1/2x + 2 as y = 1/(2x) + 2, which is a rational function, not a line.

Some people use a slash (/) to indicate division. You should use a horizontal division bar for division. A slash is okay when there is no implied multiplication in the denominator.

The TI83 properly handles the expression and all six expressions above will graph properly.

Functions

Definitions

Relation: A rule that associates a value in the domain with a value in the range.

Function: A function is a relation (rule) that assigns each element in the domain to exactly one element in the range.

Domain: The set of all the values which may be input into a function. That is, the set of all the values the independent variable may assume. Graphically, the domain is the set of all the x-coordinates.

Range: The set of all the values which are output when the function is evaluated at all the input values from the domain. That is, the set of all the values the dependent variable may assume. Graphically, the range is the set of all the y-coordinates.

Independent Variable

Typically, the independent variable is x. However, the independent variable is the variable which is free to assume different values independently of the other variable. Most of what we're going to do in this class will only involve one independent variable, but realize that it is possible to have more than one independent variable.

Dependent Variable

Typically, the dependent variable is y. However, the dependent variable is the variable which is determined based on the value of the independent variable(s). If a function is written as y=3x+2, then y depends on x, but x doesn't depend on y (in the form it's written in). So, x is independent of y, but y is dependent of x.

Implied Domain

The set of all real numbers for which the expression is defined.

How is a function different from a relation?

Here are some guidelines for determining whether a relation is a function or not.

1. Each and every element in the domain (x) must be matched with an element in the range (y).
2. Every element in the domain (x) can only be matched with one element in the range (y).
3. Different elements in the domain can go to the same element in the range. A y-coordinate may be repeated, but an x-coordinate may not.
4. Some values in the range don't have to be used at all.

Function Notation

Function notation is used to name functions for easy reference. Imagine if every function in the world had to start off with y=. Pretty soon, you would become confused about which y= you were talking about. This is the way antecedents work in English. If you just say "it" is red, you really don't have any clue which "it" you're talking about. You need some other way of naming things. Enter function notation.

Function Definition

- $f(x) = 3x + 2$
- $g(x,y) = x^2 + 3y$

In this example, the f is a function of x. That is, x is the independent variable, and the value of f depends on what x is. Also, g is a function of both x and y.

The notation f(x) does not mean f times x. It means the "value of f evaluated at x" or "value of f at x" or simply "f of x"

Function Evaluation

$$f(3) = 3(3) + 2 = 9 + 2 = 11$$

f(3) does not mean f times 3. It means the "value of f evaluated when x is 3".

$$f(t) = 3(t) + 2 = 3t + 2$$

Whatever is in parentheses on the left side of the function (t in this case) is substituted for the value of the independent variable on the right side.

$$f(x+h) = 3(x+h) + 2 = 3x + 3h + 2$$

Every occurrence of the independent variable is replaced by the quantity in parentheses. A common mistake is to take a quantity and apply linear or affine transformations to it.

$$f(x+h)$$

- $f(x+h)$ *does not equal* $f(x) + h = 3x+2 + h$
- $f(x+h)$ *does not equal* $f(x) + f(h) = 3x+2 + 3h+2 = 3x + 3h + 4$
- $f(x+h)$ does equal $3(x+h) + 2 = 3x + 3h + 2$

$$f(3x)$$

- $f(3x)$ *does not equal* $3 * f(x) = 3(3x+2) = 9x + 6$
- It does equal $3(3x) + 2 = 9x + 2$.

You also specify which function you want to use when you use function notation. Consider

$$g(2,1) = 2^2 + 3(1) = 4 + 3 = 7$$

Since the order of the independent variables in the original definition was x and then y, the function g is evaluated when x=2 and y=1.

g(1,2) is something completely different. In this case, x=1 and y=2.

$$g(1,2) = 1^2 + 3(2) = 1 + 6 = 7$$

Okay, so it's not completely different, but it is found in a different manner.

Piecewise Definitions

Sometimes, functions are a little more complex than the simple functions we've described so far. If different rules are used for different values of the independent variable, then we can use a piecewise definition.

$$f(x)=\begin{cases}3-x^2, & x<-1\\ 2x, & 1\le x<3\\ 5-x, & x\ge 3\end{cases}$$

Consider the function shown and the following evaluations.

$f(2.1) = 2(2.1) = 4.2$

Since 2.1 is in the interval [1,3), we use the second piece of the definition,

$f(x) = 2x.$

$f(-2) = 3 - (-2)^2 = 3 - 4 = -1$

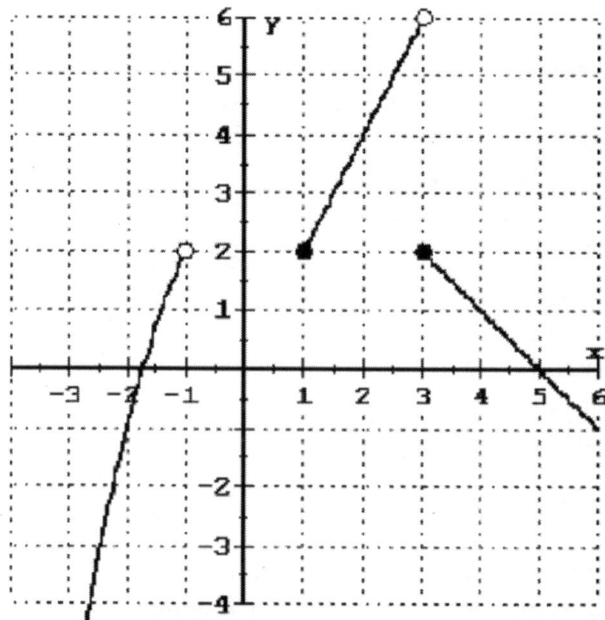

Since –2 is in the interval $(-\infty,-1)$, we use the first piece of the definition,

$f(x) = 3-x^2$

$f(3) = 5 - (3) = 5 - 3 = 2$

Since 3 is not included in the second piece, but it is included in the third piece on the interval $[3,+\infty)$ we use the third piece of the definition,

$f(x) = 5 - x.$

f(0) is undefined.

Since 0 doesn't fall into any of the domains, the function is undefined there. When all of the domains are combined, the domain of the function f, is the set of all real numbers except [–1,1). You could also write it as $(-\infty,-1)U[1,+\infty)$. The symbol U stands for the union of two sets.

Piecewise Functions and the Calculator

You can put piecewise functions into the graphing calculator. The TI82 and TI83 calculators have a [Test] key obtained by pressing [2nd] [Math]. Under this key, you will find the different test operators (equal to, not equal to, greater than, greater than or equal to, less than, and less than or equal to).

When the calculator evaluates a test expression (x<1), it will return the value of 1 if the statement is true and the value of 0 if the statement is false. This works really well with multiplication, because multiplying by 1 won't change the expression, and multiplying by 0 will make the expression 0.

$$y = (1 - x^2)*(x < -1)$$

Consider just the first part of the piecewise definition from above. If x<–1, then the calculator will return (1–x^2). If x is not less than negative one, then the calculator will return 0. That's not the proper value to return (it should be undefined), but it will graph on the x–axis and won't show up because the axis is already there. So, it has the semblance on graphing correctly.

The complete piecewise function can be defined for the calculator as:

$$y = (1-x^2)*(x<-1) + 5x*(1 \le x \text{ and } x<3) + (5-3x)*(x \ge 3)$$

The "and" keyword can be found under Logic menu. Similarly, the ≤ and e" symbols can also be found under the menu. If you try to put in 1≤x<3 like we right, the calculator will interpret that as 1≤3 and always return true.

Be sure to use some kind of decimal setting or dot mode when you're viewing a piecewise function. Otherwise, you may get some weird results.

Finding the Domain

Implied Domain

We defined *implied domain* earlier. It is the set of all real numbers where the expression is defined.

1. Start off with all real numbers
2. Exclude any values of the independent variable which cause division by zero
3. Exclude any values of the independent variable which result in taking the square root of a negative number. Square root can be replaced with any even (4th, 6th, etc) root.

You do *not* need to state restrictions which result from the implied domain. In other words, if there is an (x–2) in the denominator, you do not need to state that x cannot be 2.

Other Exclusions

Sometimes, you will need to exclude other values.

1. Problem Constraints - these are used in applications. Here are some examples
   - o If x is the length of the side of a triangle, then x can't be negative
   - o If x is the position on a 12' teeter totter at which the fulcrum must be positioned, then x must be between 0 and 12.
   - o If x is the number of people in a room, then x can't be negative.
2. Stated Restrictions - sometimes the problem will just come out and say you can't use a specific value, or that you can only use specific values. These aren't in the implied domain, so they need to be stated.

If you ever simplify a function, and a value which was in the implied domain is no longer in the implied domain, then it needs to become a stated restriction. Example, divide (x^2–4) by (x–2). The implied domain is x cannot equal 2 because that would cause division by zero. However, if you factor the numerator as (x–2)(x+2), then (x–2) in the numerator divides out with the (x–2) in the denominator and you're left with just (x+2). The fact that x cannot be 2 is no longer implied by a simple x+2 in the numerator, so you must now state that x cannot be 2.

Combining Domains

When you have a function which is a composition of several pieces, the values in the domain must be able to be used in all parts of the function. Let's say the numerator has a square root of x in it (so that x must be non-negative) and the denominator has an x–2 in it (so that x cannot be 2). When you combine those domains, you get all values of x that are non-negative except 2.

In other words, the domain for the function is the intersection of all the domains of the individual parts. Now, let's say the numerator has a square root of x in it (so that x must be non-negative) and the denominator has an x+2 in it (so that x cannot be –2). When you combine those domains, you get all values of x that are non-negative except –2. Well, you don't need to state that x cannot be -2 since you've already said that x is non-negative. So, in this case, the domain would be all non-negative values of x.

Graphs of Functions

Definitions

These definitions are mathematically loose (that means a mathematician would pull his or her hair out but a normal person might understand them).

Graph of a function

The graph of a function f is the set of all ordered pairs (x, f(x)) where x is in the domain of f.

Increasing Function

A function is increasing on an open interval if the function rises (positive slope) on the interval as you move from left to right.

Decreasing Function

A function is decreasing on an open interval if the function falls (negative slope) on the interval as you move from left to right.

Constant Function

A function is constant on an open interval if the function remains constant (horizontal line segment) on the interval as you move from left to right.

Relative Minimum

A function has a relative minimum at x=a if the function evaluated at x=a is less than at any other point in the neighbourhood surrounding x=a. A *relative* minimum is the lowest point in an open interval, but not necessarily over the entire domain. Relative minimums occur when the function is decreasing to the left of the point and increasing to the right of the point.

Relative Maximum

A function has a relative maximum at x=a if the function evaluated at x=a is greater than at any other point in the neighbourhood surrounding x=a. A *relative* maximum is the highest point in an open interval, but not necessarily over the entire domain. Relative maximums occur when the function is increasing to the left of the point and decreasing to the right of the point.

Greatest Integer Function

The greatest integer of a value is the largest integer less than or equal to the value.

Symmetry about the y-axis

A relation is symmetric about the y-axis if for every point (x,y) on the graph, the point (–x,y) is also on the graph.

Symmetry about the x-axis

A relation is symmetric about the x-axis if for every point (x,y) on the graph, the point (x,–y) is also on the graph.

Symmetry about the Origin

A relation is symmetric about the origin if for every point (x,y) on the graph, the point (–x,–y) is also on the graph.

Even Function

A function is even if for every x in the domain of the function,

$f(-x) = f(x)$

Odd Function

A function is odd if for every x in the domain of the function,

$f(-x) = -f(x)$

Vertical Line Test

A relation is a function if all vertical lines drawn through the graph of the relation intersect in no more than one point.

The contrapositive of that is often used.

If a vertical line intersects the graph of a relation in two or more points, then the relation is *not* a function.

Greatest Integer Function

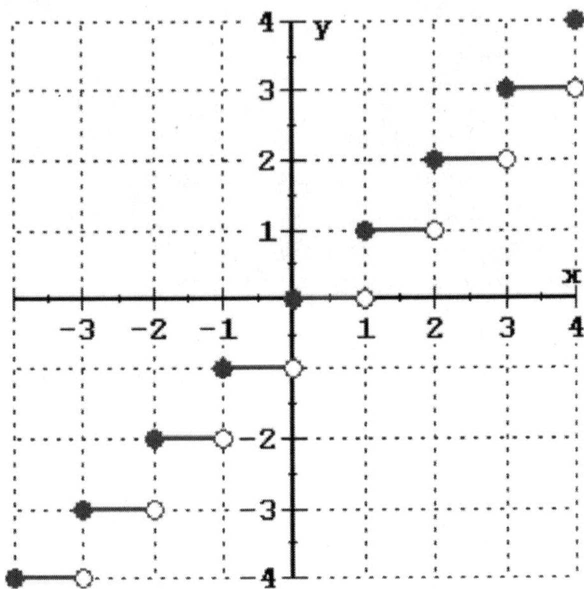

The greatest integer function is often called the Integer function (or Floor in upper level mathematics), and is abbreviated INT on the calculator. You may find the INT function on the calculator by going into the [Math] menu, arrowing right to the NUM option, and then choosing the INT function (it's number 5 on the TI83).

The Integer function is sometimes called the step function because of the stair step effect obtained when graphing it. Be sure to use a decimal setting when graphing the greatest integer function, or you will get weird results. You may also want to use Dot mode instead of Connected mode when graphing the Integer function. You can change modes on the TI series by hitting the [Mode] key.

Mathematically, the greatest integer function is represented using a double left bracket and double right bracket

Symmetry - Odd/Even Functions

Symmetry about the y-axis

Symmetry about the y-axis means that the left side of the graph is a mirror image of the right side of the graph. Mathematically, a relation which is symmetric about the y–axis has the property that for every point (x,y) that is on the graph, the point (–x,y) is also on the graph. In other words, to mirror something about the y-axis, take the opposite of all the x-coordinates and leave the y-coordinates alone.

Symmetry about the x-axis

Symmetry about the x-axis means that the bottom side of the graph is a mirror image of the top side of the graph.

Mathematically, a relation which is symmetric about the x-axis has the property that for every point (x,y) that is on the graph, the point (x,–y) is also on the graph. In other words, to mirror something about the x–axis, take the opposite of all the y-coordinates and leave the x-coordinates alone.

Symmetry about the origin

Symmetry about the origin means that for every point (x,y) on the graph, the point (–x,–y) is also on the graph.

Graphically, to create symmetry about the origin, take any point, draw an imaginary line through the origin, and place a point on that line the same distance as the original point was from the origin on the other side of the origin.

Even Function

A function is even if it is symmetric about the y–axis. Mathematically, since the y–coordinates (values of the function) have to be equal, and the x–coordinates are opposite, one can write: $f(-x) = f(x)$.

There is a reason these functions are called even. If you have a polynomial function in one variable, all of the exponents on the independent variable will be even. Remember that a constant is the zero power of the variable and zero is even.

Odd Function

A function is odd if it is symmetric about the origin. Mathematically, since both the x–coordinate and the y-coordinates are negated, one can write:

$$f(-x) = -f(x).$$

There is a reason these functions are called odd. If you have a polynomial function in one variable, all of the exponents on the independent variable will be odd. Remember that a constant is the zero power of a variable and will be even, so if there is a constant term, it is not an odd function.

Shifting, Reflecting, and Stretching Graphs

Definitions

Abscissa: The x–coordinate

Ordinate: The y-coordinate

Shift: A translation in which the size and shape of a graph of a function is not changed, but the location of the graph is.

Scale: A translation in which the size and shape of the graph of a function is changed.

Reflection: A translation in which the graph of a function is mirrored about an axis.

Common Functions

Part of the beauty of mathematics is that almost everything builds upon something else, and if you can understand the foundations, then you can apply new elements to old.

It is this ability which makes comprehension of mathematics possible.

If you were to memorize every piece of mathematics presented to you without making the connection to other parts, you will

1) become frustrated at math and
2) not really understand math.

There are some basic graphs that we have seen before. By applying translations to these basic graphs, we are able to obtain new graphs that still have all the properties of the old ones.

By understanding the basic graphs and the way translations apply to them, we will recognize each new graph as a small variation in an old one, not as a completely different graph that we have never seen before.

Understanding these translations will allow us to quickly recognize and sketch a new function without having to resort to plotting points.

Translations

There are two kinds of translations that we can do to a graph of a function. They are shifting and scaling.

There are three if you count reflections, but reflections are just a special case of the second translation.

Shifts

A shift is a rigid translation in that it does not change the shape or size of the graph of the function. All that a shift will do is change the location of the graph.

A vertical shift adds/subtracts a constant to/from every y–coordinate while leaving the x-coordinate unchanged.

A horizontal shift adds/subtracts a constant to/from every x–coordinate while leaving the y–coordinate unchanged. Vertical and horizontal shifts can be combined into one expression.

Shifts are added/subtracted to the x or f(x) components. If the constant is grouped with the x, then it is a horizontal shift, otherwise it is a vertical shift.

Scales (Stretch/Compress)

A scale is a non–rigid translation in that it does alter the shape and size of the graph of the function.

A scale will multiply/divide coordinates and this will change the appearance as well as the location.

A vertical scaling multiplies/divides every y–coordinate by a constant while leaving the x–coordinate unchanged.

A horizontal scaling multiplies/divides every x–coordinate by a constant while leaving the y–coordinate unchanged. The vertical and horizontal scalings can be combined into one expression.

Scaling factors are multiplied/divided by the x or f(x) components. If the constant is grouped with the x, then it is a horizontal scaling, otherwise it is a vertical scaling.

Reflections

A function can be reflected about an axis by multiplying by negative one. To reflect about the y–axis, multiply every x by –1 to get –x.

To reflect about the x–axis, multiply f(x) by –1 to get –f(x).

Putting it all Together

Consider the basic graph of the function: y = f(x)

All of the translations can be expressed in the form:

$$y = a * f[b(x-c)] + d$$

	Vertical	*Horizontal*
Scale	a	b
Shift	d	c
	acts normally	acts inversely

Digression

Understanding the concepts here are fundamental to understanding polynomial and rational functions and especially conic sections. It will also play a very big roll in Trigonometry and Calculus.

Earlier in the text, there were a series of problems which wrote the equation of a line as:

$$x/a + y/b = 1$$

Where a was the x-intercept and b was the y-intercept of the line. The "a" could really be thought of how far to go in the x-direction (an x-scaling) and the "b" could be thought of as how far to go in the "y" direction (a y-scaling). So the "a" and "b" there are actually multipliers (even though they appear on the bottom). What they are multiplying is the 1 which is on the right side. x+y=1 would have an x-intercept and y-intercept of 1.

Okay. Consider the equation: y = f(x)

This is the most basic graph of the function. But transformations can be applied to it, too. It can be written in the format shown to the below.

$$\frac{y-d}{a} = f\left(\frac{x-c}{b}\right)$$

In this format, the "a" is a vertical multiplier and the "b" is a horizontal multiplier. We know that "a" affects the y because it is grouped with the y and the "b" affects the x because it is grouped with the x.

The "d" and "c" are vertical and horizontal shifts, respectively. We know that they are shifts because they are subtracted from the variable rather than being divided into the variable, which would make them scales.

In this format, all changes seem to be the opposite of what you would expect. If you have the expression (y–2)/3, it is a vertical shift of 2 to the right (even though it says y *minus* 2) and it is a vertical stretching by 3 (even though it says y *divided by* 3). It is important to realize that in this format, when the constants are grouped with the variable they are affecting, the translation is the opposite (inverse) of what most people think it should be.

However, this format is not conducive to sketching with technology, because we like functions to be written as y =, rather than (y-c)/d =. So, if you take the notation above and solve it for y, you get the notation below, which is similar, but not exactly our basic form state above.

y = a * f((x–c) / b) + d

Note that to solve for y, you have had to inverse both the "a" and "d" constants. Instead of dividing by "a", you are now multiplying by "a". Well, it used to be that you had to apply the inverse of the constant anyway. When it said "divide by a", you knew that it meant to "multiply each y by a". When it said "subtract d", you knew that you really had to "add d". You have already applied the inverse, so don't do it again! With the constants affecting the y, since they have been moved to the other side, take them at face value. If it says multiply by 2, do it, don't divide by 2.

However, the constants affecting the x have not been changed. They are still the opposite of what you think they should be. And, to make matters worse, the "x divided b" that really means multiply each x-coordinate by "b" has been reversed to be written as "b times x" so that it really means divide each x by "b". The "x minus c" really means add c to each x-coordinate.

So, the final form (for technology) is as above:

y = a * f [b (x-c)] + d

Ok, end of digression.

Normal & Inverse Behaviour: You will notice that the chart says the vertical translations are normal and the horizontal translations are inversed. For an explanation of why, read the digression above. The concepts in there really are fundamental to understanding a lot of graphing.

Examples

y=f(x)

No translation

y=f(x+2)

The +2 is grouped with the x, therefore it is a horizontal translation. Since it is added to the x, rather than multiplied by the x, it is a shift and not a scale. Since it says *plus* and the horizontal changes are inversed, the actual translation is to move the entire graph to the left two units or "subtract two from every x-coordinate" while leaving the y-coordinates alone.

$$y=f(x)+2$$

The +2 is not grouped with the x, therefore it is a vertical translation. Since it is added, rather than multiplied, it is a shift and not a scale. Since it says *plus* and the vertical changes act the way they look, the actual translation is to move the entire graph two units up or "add two to every y-coordinate" while leaving the x-coordinates alone.

$$y=f(x-3)+5$$

This time, there is a horizontal shift of three to the right and vertical shift of five up. So the translation would be to move the entire graph right three and up five or "add three to every x-coordinate and five to every y-coordinate"

$$y=3f(x)$$

The 3 is multiplied so it is a scaling rather than a shifting. The 3 is not grouped with the x, so it is a vertical scaling. Vertical changes are affected the way you think they should be, so the result is to "multiply every y-coordinate by three" while leaving the x-coordinates alone.

$$y=-f(x)$$

The y is to be multiplied by –1. This makes the translation to be "reflect about the x–axis" while leaving the x-coordinates alone.

$$y=f(2x)$$

The 2 is multiplied rather than added, so it is a scaling instead of a shifting. The 2 is grouped with the x, so it is a horizontal scaling. Horizontal changes are the inverse of what they appear to be so instead of multiplying every x-coordinate by two, the translation is to "divide every x-coordinate by two" while leaving the y-coordinates unchanged.

$$y=f(-x)$$

The x is to be multiplied by –1. This makes the translation to be "reflect about the y–axis" while leaving the y–coordinates alone.

$$y=1/2\ f(x/3)$$

The translation here would be to "multiply every y-coordinate by 1/2 and multiply every x-coordinate by 3".

y=2f(x)+5

There could be some ambiguity here. Do you "add five to every y-coordinate and then multiply by two" or do you "multiply every y–coordinate by two and then add five"?

This is where my comment earlier about mathematics building upon itself comes into play. There is an order of operations which says that multiplication and division is performed before addition and subtraction. If you remember this, then the decision is easy.

The correct transformation is to "multiply every y-coordinate by two and then add five" while leaving the x-coordinates alone.

y=f(2x–3)

Now that the order of operations is clearly defined, the ambiguity here about which should be done first is removed.

The answer is not to "divide each x–coordinate by two and add three" as you might expect. The reason is that problem is *not* written in standard form. Standard form is y=f[b(x–c)]. When written in standard form, this problem becomes y=f[2(x–3/2)].

This means that the proper translation is to "divide every x–coordinate by two and add three-halves" while leaving the y-coordinates unchanged.

y=3f(x–2)

The translation here is to "multiply every y-coordinate by three and add two to every x–coordinate". Alternatively, you could change the order around. Changes to the x or y can be made independently of each other, but if there are scales and shifts to the same variable, it is important to do the scaling first and the shifting second.

Translations and the Effect on Domain & Range

Any horizontal translation will affect the domain and leave the range unchanged. Any vertical translation will affect the range and the leave the domain unchanged.

Apply the same translation to the domain or range that you apply to the x-coordinates or the y-coordinates. This works because the domain can be written in interval notation as the interval between two x-coordinates. Likewise for the range as the interval between two y-coordinates.

In the following table, remember that domain and range are given in interval notation. If you're not familiar with interval notation. The first line is the definition statement and should be used to determine the rest of the answers.

Graph	*Translation*	*Domain*	*Range*
y=f(x)	none	(–2,5)	[4,8]
y=f(x–2)	right 2	(0,7)	[4,8]
y=f(x)–2	down 2	(–2,5)	[2,6]
y=3f(x)	multiply each y by 3	(–2,5)	[12,24]
y=f(3x)	divide each x by 3	(–2/3,5/3)	[4,8]
y=2f(x–3)–5	multiply each y by 2 and subtract 5; add 3 to every x	(1,8)	[3,11]
y=–f(x)	reflect about x-axis	(–2,5)	[–8,–4]
y=1/f(x)	take the reciprocal of each y	(–2,5)	[1/8,1/4]

Notice on the last two that the order in the range has changed. This is because in interval notation, the smaller number always comes first.

Really Good Stuff

Understanding the translations can also help when finding the domain and range of a function. Let's say your problem is to find the domain and range of the function y=2-sqrt(x–3).

Begin with what you know. You know the basic function is the sqrt(x) and you know the domain and range of the sqrt(x) are both [0,+infinity). You know this because you know those six common functions on the front cover of your text which are going to be used as building blocks for other functions.

	Function	*Translation*	*Domain*	*Range*
Begin with what you know	y=sqrt(x)	None	[0,+infinity)	[0,+ infinity)
Apply the translations	y=-sqrt(x)	Reflect about x-axis	[0,+infinity)	(-infinity,0]
	y=2-sqrt(x)	Add 2 to each ordinate	[0,+infinity)	(–infinity,2]
	y=2-sqrt(x–3)	Add 3 to each abscissa	[3,+infinity)	(–infinity,2]

So, for the function y=2-sqrt(x-3), the domain is $x \geq 3$ and the range is $y \leq 2$.

And the *best part* of it is that you understood it! Not only did you understand it, but you were able to do it without graphing it on the calculator.

There is nothing wrong with making a graph to see what's going on, but you should be able to understand what's going on without the graph because we have learned that the graphing calculator doesn't always show exactly what's going on. It is a tool to assist your understanding and comprehension, not a tool to replace it. It is this cohesiveness of math that I want all of you to "get". It all fits together so beautifully.

Combinations of Functions

Arithmetic Combinations of Functions: The sum, difference, product, or quotient of functions can be found easily.

Sum

$$(f + g)(x) = f(x) + g(x)$$

Difference

$$(f - g)(x) = f(x) - g(x)$$

Product

$$(f \cdot g)(x) = f(x) \cdot g(x)$$

Quotient

$(f / g)(x) = f(x) / g(x)$, as long as g(x) isn't zero.

The domain of each of these combinations is the intersection of the domain of f and the domain of g. In other words, both functions must be defined at a point for the combination to be defined. One additional requirement for the division of functions is that the denominator can't be zero, but we knew that because it's part of the implied domain.

Basically what the above says is that to evaluate a combination of functions, you may combine the functions and then evaluate or you may evaluate each function and then combine.

Composition of Functions

While the arithmetic combinations of functions are straightforward and fairly easy, there is another type of combination called a composition.

A composition of functions is the applying of one function to another function. The symbol of composition of functions is a small circle between the function names. I can't do that symbol in text mode on the web, so I'll use a lower case oh "o" to represent composition of functions.

- (fog)(x) = f [g(x)]
- (gof)(x) = g [f(x)]

These are read "f composed with g of x" and "g composed with f of x" respectively.

The function on the outside is always written first with the functions that follow being on the inside. The order is important. Composition of functions is not commutative.

Examples of Composition of Functions: That doesn't sound that bad. Let's look at a few examples.

$f(x)=5x+2$ and $g(x)=x^2-1$

- $(fog)(x) = f[g(x)] = f[x^2-1] = 5(x^2-1) + 2 = 5x^2- 5 + 2 = 5x^2-3$
- $(gof)(x) = g[f(x)] = g[5x+2] = (5x+2)^2-1 = 25x^2 + 20x + 4-1 = 25x^2 + 20x + 3$

f(x) = sqrt(x) and $g(x) = 4x^2$

- (fog)(x) = f [g(x)] = f [4x2] = sqrt($4x^2$) = 2 | x |
- (gof)(x) = g [f(x)] = g [sqrt(x)] = 4 (sqrt(x))2 = 4x, $x \geq 0$

This example probably needs some explanation. From the prerequisite chapter, the square root of (x^2) is the absolute value of x. The square of (the square root of x) is x, but this assumes that x is not negative because you couldn't find the square root of x in the first place if it was. This is a case where the implied domain is no longer implied, so you have to explicitly state it.

f(x) = sqrt(x–4) and $g(x) = 1 - x^2$

- (fog)(x) = f [g(x)] = f [$1-x^2$] = sqrt ([$1-x^2$] – 4) = sqrt ($-x^2 - 3$) = ψ
- (gof)(x) = g [f(x)] = g [sqrt(x-4)] = 1 – $[\text{sqrt}(x-4)]^2$ = 1– (x–4) = 5 – x, $x \geq 4$

If the last example needed some explanation, then this one definitely needs some, too. Let's take the easier one (gof)(x) first. There was an implied domain of $x \geq 4$ because of the square root, but after squaring it, it was no longer implied, so it needed to be stated explicitly.

Okay, now for the harder one (fog)(x). I'll give the simple explanation here and the more complete one later. After simplifying, you got the square root of $(-x^2 - 3)$. $-x^2-3$ is always negative, no matter what real number x is, and you can't take the square root of a negative number, so it is always undefined (for the set of reals).

Finding Domains on Composition of Functions

When you find a composition of a functions, it is no longer x that is being plugged into the outer function, it is the inner function evaluated at x. So there are two domains that we have to be concerned about. If we consider (fog)(x), we see that g is evaluated at x, so x has to be in the domain of g. We also see that f is evaluated at g(x), so g(x) has to be in the domain of f.

- For (fog)(x), x is a value that can be plugged into g and gives you a value g(x) that can be plugged into f to get f(g(x))
- For (gof)(x), x is a value that can be plugged into f and gives you a value f(x) that can be plugged into g to get g(f(x)).

But, it's not as bad as it looks, either. Let's consider that last example again.

	Function	*Domain*	*Range*
f(x) =	sqrt(x-4)	x^3 4	y^3 0
g(x) =	$1-x^2$	all reals	y £ 1

When you find (fog)(x), there are two things that must be satisfied:

1. x must be in the domain of g, which means x is a real number (pretty easy to do)
2. g(x) must be in the domain of f, which means that $1-x^{2}$^2 $\geq$ 4 (when you try to solve this, you get the empty set)

When you combine the two domains to see what they have in common, you find the intersection of everything and nothing is nothing (the empty set), so the function is defined nowhere and undefined everywhere.

When you find (gof)(x), there are two things that must be satisfied:

1. x must be in the domain of f, which means that $x \geq 4$ (not too bad)
2. f(x) must be in the domain of g which means that the sqrt(x–4) must be a real number (that occurs when $x \geq 4$, which we already have stated from the first part)

When you combine the two domains to see what they have in common, you find the intersection to be $x \geq 4$, so that is the where the composition is defined.

Decomposition of Functions

Decomposition of functions is the reverse of composition of functions. Instead of combining two functions to get a new function, you're breaking apart a combined function into its separate components. There is often more than one way to decompose a function, so your answers may vary from the books.

Basically, you want to look at the function and look for an "outside function" and an "inside function". Another thing to look for is repeated patterns and make that the inside function. The outside function is summarized as "the big picture" and the inside function is "what you are doing the big picture to".

Difference Quotients

Difference quotients are what they say they are. They involve a difference and a quotient. A difference quotient is really the slope of a secant line between two points on a curve.

The formula for the difference quotient is: $[f(x+h)-f(x)]/h$

If you go on to calculus, you will be introduced to the concept of limits (Calculus is Algebra with limits) and find as the two points on the curve get closer together, the secant line turns into a tangent line, and the slope of the secant line becomes the slope of the tangent line which is called the derivative of the function and there are a whole lot of shortcuts for finding derivatives. However, you're still in College Algebra and don't know about limits, so you'll have to find the difference quotients by hand. The difference quotient is very important in Calculus, so if you are going on, make sure you get this!

For polynomial functions, finding the difference quotient isn't that difficult. Where you're going to run into trouble is with radical and rational functions.

Polynomial Functions

$f(x) = 5x^2 - 2$

$f(x+h) = 5(x+h)^2 - 2 = 5(x^2 + 2xh + h^2) - 2 = 5x^2 + 10xh + 5h^2 - 2$

$f(x+h) - f(x) = 5(x^2 + 2xh + h^2) - 2 = 5x^2 + 10xh + 5h^2 - 2 - (5x^2 - 2)$

$f(x+h) - f(x) = 5x^2 + 10xh + 5h^2 - 2 - 5x^2 + 2 = 10xh + 5h^2$

$f(x+h) - f(x) = h\ (10x + 5h)$

$[f(x+h) - f(x)]\ /\ h = h\ (10x + 5h)/h = 10x + 5h$

Rational Functions

The trick with rational functions is to get a common denominator and then simplify.

$f(x) = 3\ /\ x$

$f(x+h) = 3\ /\ (x+h)$

$f(x+h) - f(x) = 3/(x+h) - 3/x$

$f(x+h) - f(x) = 3\ x\ /\ [x(x+h)] - 3(x+h)/[x(x+h)]$

$f(x+h) - f(x) = (3x - 3x - 3h)/[x(x+h)]$

$f(x+h) - f(x) = -3h\ /\ [x(x+h)]$

$[f(x+h) - f(x)]/h = -3h/[x(x+h)]/h = -3/[x(x+h)]$

Inverse Functions

Notation

The inverse of the function f is denoted by f^{-1} (if your browser doesn't support superscripts, that is looks like f with an exponent of –1) and is pronounced "f inverse". Although the inverse of a function looks like you're raising the function to the –1 power, it isn't. The inverse of a function does not mean the reciprocal of a function.

Inverses

A function normally tells you what y is if you know what x is. The inverse of a function will tell you what x had to be to get that value of y.

A function f^{-1} is the inverse of f if

- for every x in the domain of f, $f^{-1}[f(x)] = x$, and
- for every x in the domain of f^{-1}, $f[f^{-1}(x)] = x$

The domain of f is the range of f^{-1} and the range of f is the domain of f^{-1}.

Graph of the Inverse Function

The inverse of a function differs from the function in that all the x-coordinates and y-coordinates have been switched. That is, if (4,6) is a point on the graph of the function, then (6,4) is a point on the graph of the inverse function.

Points on the identity function (y=x) will remain on the identity function when switched. All other points will have their coordinates switched and move locations.

The graph of a function and its inverse are mirror images of each other. They are reflected about the identity function y=x.

Existence of an Inverse Function

A function says that for every x, there is exactly one y. That is, y values can be duplicated but x values can not be repeated.

If the function has an inverse that is also a function, then there can only be one y for every x.

A one-to-one function, is a function in which for every x there is exactly one y and for every y, there is exactly one x.

A one-to-one function has an inverse that is also a function.

There are functions which have inverses that are not functions. There are also inverses for relations. For the most part, we disregard these, and deal only with functions whose inverses are also functions.

If the inverse of a function is also a function, then the inverse relation must pass a vertical line test. Since all the x-coordinates and y-coordinates are switched when finding the inverse, saying that the inverse must pass a vertical line test is the same as saying the original function must pass a horizontal line test.

If a function passes both the *vertical line test* (so that it is a function in the first place) and the *horizontal line test* (so that its inverse is a function), then the function is one-to-one and has an inverse function.

Finding Inverses Informally

The inverse of some functions, especially those where there is only one occurrence of the independent variable, can be solved by *undoing* the operations. To undo the operations, you must not only reverse the order, but use the inverse operation.

Example 1

The function f(x) = 5x–2

1. Start with x: x
2. Multiply by 5: 5x
3. Subtract 2: 5x-2

The inverse $f^{-1}(x) = (x+2)/5$

1. Start with x: x
2. Add 2: x+2
3. Divide by 5: (x+2)/5

Example 2

The function $f(x) = 2(x-3)^2-5,\ x^{\geq}3$

Note that there is a restriction on x.

1. Start with x: x
2. Subtract 3: x–3
3. Square: $(x-3)^2$
4. Multiply by 2: $2(x-3)^2$
5. Subtract 5: $2(x-3)^2-5$

The inverse $f^{-1}(x) = 3 + \text{sqrt}[(x+5)/2]$

1. Start with x: x
2. Add 5: x + 5
3. Divide by 2: (x+5)/2
4. Take the square root: ± sqrt [(x+5)/2]
5. Add 3: 3 ± sqrt[(x+5)/2]
6. Wait! That inverse isn't a function because there are two values of y for every x. That's because of the ± that appeared when we took the square root of both sides. Now we go back to the original domain of x≥3. T

   hat means that for the inverse, the range is y≥3. Since y must be at least 3, we need the positive square root and not the negative.

   Without the restriction on x in the original function, it wouldn't have had an inverse function: 3 + sqrt[(x+5)/2]

Example 3

The function $f(x) = x^2 - 4x + 6,\ x \leq 2$

What happens when there is more than one occurrence of the independent variable in the function? You don't know what you did to x because you did it to two different x's and you didn't do the same thing to both of them.

Finding Inverses Formally

Now, that last example is not to be said it can't be done, but it involves completing the square to obtain $f(x) = (x–2)^2+2$, then inversing it so that you get $f^{-1}(x) = 2–\text{sqrt}(x–2)$. However, there is another way that doesn't rely so much on informality and will work whether or not you can figure out exactly what you did with exactly one x.

1. Start with the function
2. Replace f(x) by y if necessary
3. Switch the x's and y's. At this point you are dealing with the inverse
4. Solve for y
5. Replace y with $f^{û1}(x)$ if the inverse is also a function, otherwise leave it as y.

Example 4

The function $f(x) = x^2/(x^2+1),\ x \geq 0$

The restriction is important to make it 1–1.

1. Start with the function: $f(x) = x^2 / (x^2+1),\ x \geq 0$
2. Replace f(x) by y: $y = x^2/(x^2+1),\ x \geq 0$
3. Switch the x's and y's: $x = y^2 / (y^2+1),\ y \geq 0$
4. Solve for y:
    1. Multiply by denominator: $x(y^2+1) = y^2$
    2. Distribute: $xy^2 + x = y^2$
    3. Move y's to one side and everything else to other: $xy^2–y^2 = –x$
    4. Factor: $y^2(x–1)=–x$
    5. Divide by coefficient on y^2: $y^2 = –x/(x–1)$
    6. Simplify right hand side: $y^2 = x/(1–x)$
    7. Take square root: $y = \pm\ \text{sqrt}[\ x/(1–x)\ ]$
    8. Since $y \geq 0$, we need the positive square root: $y = \text{sqrt}[x/(1–x)]$
5. Call this $f^{-1}(x)$: $f^{-1}(x) = \text{sqrt}[x/(1–x)]$

On this last function, the implied domain of the inverse is [0,1). That means that the range of the original function must have been [0,1), also. Check it on your calculator, and you'll see it is.

Sometimes the instructions say if the function is not one-to-one, then don't find the inverse function (because there's not one). So, always check before wasting time trying to find the inverse function. Now, if you're supposed to find the inverse, regardless of whether it is a function or not, then go ahead.

Good Stuff!

One-to-One Functions are wonderful things.

When solving equations, you can add the same thing to both sides, subtract the same thing from both sides, multiply both sides by the same non-zero thing, and divide both sides by the same non-zero thing and still get the same solution without worrying about having to check your answer.

You can also apply a one-to-one function to both sides of an equation without worrying about introducing extraneous solutions (solutions that work after doing something that didn't work before). This is not necessarily true with functions that aren't one-to-one like the squaring function where you should always check answers after you square both sides of an equation.

For example, the equation sqrt(x) = –2 has no solution, but if you square both sides, you get x = 4, but it doesn't check in the original problem. With one-to-one functions, you won't be introducing any extraneous solutions.

Wow! Talk about powerful. You don't appreciate it now, and the book doesn't deal with it properly until and deal with logarithmic and exponential functions, and even then they don't make as big of deal out of it as it is.

Okay, let's try one now. Take my word for it that exp(x) is a one-to-one function and is the inverse of ln(x).

$$\ln(x) = 3$$

Solve for x.

$$\exp[\ln(x)] = \exp\ [3]$$

"Wait a minute, Mr. Jones" is your response. You've never seen such a beast. That's okay. Take the inverse of the function, and apply it to both sides.

$$x = \exp\ (3)$$

Go back to the definition of an inverse at the top of this document. When inverses are applied to each other, they inverse each other out, and you're just left with the argument (input) to the function.

$x = e^3$

On the calculator, the exp(x) function is written e^x, and is found on the [2nd] [ln] key.

Wow - more cohesiveness. The inverse of a function is found by taking the [2nd] function. Look at it for other things on the calculator.

The square root is the inverse of the square. If you look at the three trigonometric keys [sin], [cos], and [tan], their inverses are all found by using the [2nd] key.

Soapbox mode on.

I'm telling you - it all fits together. For those who remember the line Hannibal Smith used in the A-Team, "I love it when a plan comes together".

Mathematics is one of the most together subjects there is. Everything complements everything else. What I'm hoping you get out of this course is much more than just the mechanics of mathematics, but a comprehension, understanding, and appreciation of the way the system works. With that solid foundation, mathematics can be less stressful, and yes, even enjoyable. You have to stop dealing with concepts as individual things that don't relate to each other and stand alone. They are all related to each other and intertwined. You can not separate them and understand.

Bibliography

Aaboe, Asger: *Episodes from the Early History of Mathematics*, New York, 1964.

Arthur W. Gilbert, Mortier F. Barrus and Daniel Dean: *Growing and Breeding of Potatoes*, Asiatic Pub, Delhi, 2006.

Barbeau G.: *Tropical Fruits in Nicaragua*, Managua, Nicaragua Ministerio de Desarrollo Agropecuario, Agraria, 1990.

Batlle I. and Tous J.: *Carob Tree (Ceratonia siliqua L.)*. Rome, International Plant Genetic Resources Institute, 1997.

Collymore L.: *Fruit Production in Barbados*, Port of Spain, Trinidad and Tobago, 1996.

Cronquist, A.: *The Evolution and Classification of Flowering Plants*, New York Botanical Garden, Bronx, New York., 1988

Dodia; DA.: IS Patel and GM Patel: *Botanical Pesticides for Pest Management*, Scientific, Delhi, 2008.

Daniells J.: *Illustrated Guide to the Identification of Banana varieties in the South Pacific*, Canberra, ACIAR, 1995.

Ferentinos L.: *Proceeding of the Sustainable Taro Culture for the Pacific Conference*, Honolulu, HITAHR, 1993.

Fortuner, R.: *Nematode Identification and Expert System Technology*, New York, Plenum Press, 1988.

Georges, S.: *The Debt Boomerang: How Third World Debt Harms Us All*, Boulder, Westview Press, 1992.

Godden G.: *Growing Citrus Trees*, Australia, Lothian Publishing Company Pvt. Ltd., 1988.

Hardy B.: *Biology and Agronomy of Forage Arachis*, Cali, International Centre for Tropical Agriculture, 1994.

Jeffers P.: *Evaluation of Four Onion Varieties in Montserrat*, Plymouth, CARDI, 1992.

Jones, R. M.: *Plant Resources of South-East Asia*, Wageningen, Pudoc Scientific Publishers, 1992.

Kim J. S.: *Sources of Resistance to Viruses of Pepper (Capsicum spp.): a Catalogue*, Taipei, AVRDC, 1994.

Kuppuram, G & K. Kumudamani: *History of Science and Technology in India*, Delhi, Sundeep Prakashan, 1990.

Lorenzen, S.: *The Phylogenetic Systematics of Freeliving Nematodes*, London, The Ray Society, 1994.

Madulid Domingo A.: *A Pictorial Cyclopedia of Philippine Ornamental Plants*, Philippines, Makati Metro Manila, 1995.

Malins A.: *Postharvest Handling of Pineapple and Mango*, Port of Spain, Trinidad and Tobago, 1992.

Nobel, P. S.: *Physicochemical and Environmental Plant Physiology*, Academic Press, San Diego, 1999.

Oka H. I.: *Origin of Cultivated Rice*, Elsevier, Japan Scientific Societies Press, 1988.

Oldham P.: *Cost of Production of Major Tree Crops in Dominica*, Roseau, Ministry of Agriculture, 1991.

Parry M.L.: *Climatic Change, Agriculture and Settlements*, Dawson Folkestone UK, 1978.

Pemberton, R. W.: *Predictable Risk to Native Plants in Weed Biological Control,* Oecologia, 2000.

Qystein V. Sjaastad: *Physiology of Domestic Animals*, International Book Distributing Co., Delhi, 2005.

Rao, S. Balaachandra: *Indian Mathematics and Astronomy*, Bangalore, Jnana Deep Publications, 1994.

Sanchez, P. A.: *Properties and Management of Soils in the Tropics*, John Wiley and Sons, New York, 1976.

Sinha G. C., Reddy Y. T. N. and Singh G.: *Propagation and Rootstocks in Guava*, New Delhi, Malhotra Publishers, 1993.

T. Pullaiah & G.N.V. Febulaus: *Embryology and Apomixis in Grasses*, Regency, Delhi, 2000.

Urton, Gary: *The Social Life of Numbers*, Austin, University of Texas Press, 1997.

Vanderplank, J.E.: *Plant Diseases: Epidemics and Control,* New York, NY, USA, Academic Press, 1963.

White, G.F.: *Natural Hazards: Local, National, Global*, Oxford University Press, New York, 1974.

Woolfe Jennifer A.: *The Potato in the Human Diet*, Cambridge, Cambridge University Press, 1989.

Yoshida T.: *Cultivation of Citrus Genetic Resources for Evaluation of Characteristics*, Tokyo, JICA, 1996.

Index

L

M

N

O

P

R

S

T

U

V

W

Z

❑❑❑